To Conrad, Emma & Robert

from a shamefully neglectful godparent

this volume is humbly dedicated

in the hope that if read it will inform

and if not read

that at least it will enhance some space in a bookcase

.

First published by KDP September 2023
Minor revisions July 2024

There are known knowns — there are things we know we know. We also know there are known unknowns — that is to say, we know there are some things we do not know. But there are also unknown unknowns, the ones we don't know we don't know.

Donal Rumsfeld

Lady Bracknell	*Do you know everything or nothing?*
Jack	*I know nothing, Lady Bracknell.*
Lady Bracknell	*I am pleased to hear it. I do not approve of anything that tampers with natural ignorance.*

Oscar Wilde

PARTIAL KNOWLEDGE

SEVERAL AREAS OF IGNORANCE

FOREWORD

The question of what is meant by the verb 'to know,' has puzzled me nearly all my life. It started seventy years ago when I was still at school and then more formally whilst studying mediaeval philosophy as a Jesuit scholastic at Heythrop College, under the aegis of the Pontifical Gregorian University of Rome. It has been continued sporadically during a non-academic working life as a civil engineer. This enquiry encompasses the means by which we, and other life forms, acquire knowledge, its various forms and the extent to which credence can be placed in any of them.

That we must have some sure knowledge is prerequisite of any meaningful discussion. The statement, "There is no sure knowledge," contradicts itself if it is to have any meaning; if it is not to have any content there is not much point in making it. One might as well speak in an uncoordinated series of grunts, write in a wholly undecipherable script. No attempt will be made to discover any single matter about which we may be sure: not universal doubt but a certain scepticism where there is any room for doubt and a realisation that our knowledge is limited and occasionally mistaken.

There will be no appeals to authority whether from classical or modern philosophy, or Judaic, Christian, Islamic, Buddhist, Hindu or any other religion: only to the readers' own experiences, the current consensus of scientific knowledge and common sense. As far as is possible an attempt will be made to separate scientific fact from theory. Relativity and evolution, for example, are both partly factual and partly theoretical. This is not to say that no account has been taken of the immense literature in all these fields, far from it.

This essay is not directed to those who have had or believe they have had some mystical or extra-sensual experience. Yet, such experiences are not to be dismissed out of hand.

No firm conclusions are drawn, although one is suggested, but it is hoped that some of the loose thinking which surrounds the topic will be dispelled and that at the very least it will stimulate thought.

My thanks are due to my son Christopher for reading the manuscript and making many helpful suggestions in addition to finding many typos.

CONTENTS

Introduction

The current state of understanding of the basic constituents of matter.

The atomic composition of seemingly simple static objects; the vast numbers involved and the fact that all are in in constant motion.

With two possible exceptions everything of which we are aware comes via our five senses. These have resulted as products of evolution.

The exceptions are internal aches, pains and pleasures which have similar neural paths to the brain. There are also mystical experiences. These are not discussed here (but their possibility not dismissed either).

The extraordinary complexity of the brain. Memory: what it must entail and how little is known of its mechanism.

A brief description of the theory of evolution, leading to the enormous complexity of living organisms. The scale of these.

The unimaginable complexity of the brain and the huge number of cells in it, all in constant motion. Internal processes. The nature of memory. The material element in abstract thought. The possible illusory nature of perception.

PART 2 THE PSYCHE: Knowing living creatures.

1. Animals.

Similarities of knowing animals and knowing persons. Are they identical? Gradual increase in animal consciousness. Animal pain and anthropomorphism. Possibility of self-conscious awareness in animals. Cruelty.

2. Persons.

2.1. Self-knowledge.

Limitation; proclivity to self-delusion.

2.1.1. Motives and self-judgement.

Sometimes too lenient, sometimes too harsh. Introspection.

2.1.2. Freedom of action.

Universal belief in individual freedom: impossibility of judging whether any action is free.

2.2. Other persons.

Is knowing others the same as knowing self? Is direct knowledge of another possible?

3. Abstract knowledge.

Is the power of reasoning possible for a computer or is it uniquely human? An unsatisfactory chapter!

PART 3 VAST AREAS OF IGNORANCE

1. Knowledge from nurture: home and school.

Huge influence of knowledge gained in the nursery, the family and school. Its reliability.

2. Media based knowledge

2.1. Social media and fake news.

Print media and web-based articles.

2.2. Printed material and web-based articles.

Reliability of scientific theory. Vaccines and freedom. Smart motorways.

PART 4 RIGHT AND WRONG

1 Morality

Do humans have an innate sense of right and wrong?
Arguments from introspection and evolution. The role
of parental guidance.
Free will, its existence, nature and place in
responsibility.
Are there any absolutes when it comes to right and
wrong? How can one distinguish one from the other?

1.1 Conscience
1.1.1 Innate moral sense
1.1.2 Education
1.1.3 Responsibility

2   **Religion**
So many religions. What possible criterion
for judging which if any is true?
Is religious belief God-given or a matter of
indoctrination, particularly parental?
Arguments for the existence of a non-material God: Plato,
Aristotle and Aquinas.
Universal beliefs of nature of God in major religions
Is evil positive or only a negation of good?

PART 5 AFTER DEATH: A SPECULATION
1   The notion of ensoulment
    1.1   Spirit-matter interaction
    1.2   Spirit and memory
    1.3   Memory after death: non-material memory.
    Theories of bond between body and soul: unitary
    and dualistic. Human form determined by DNA.
2   Necessity of the Resurrection of the body.
A speculation concerning the exercise of post-mortem
memory.

**PART 6   CONCLUSION: Some certain knowledge – a
comforting speculation.**

Arcane philosophy aside, everyone knows that what is sensed is what is actually there. The tree really is a tree. The sun really rises and sets. There is a theory, necessarily theistic, which reconciles sensation with science.

6.1 Materialism

6.2 Idealism

6.3 A theistic solution: A hint of how certain knowledge might be possible in all of the above categories.

INTRODUCTION

What a piece of work is man, how noble in reason, how infinite in faculties, how like an angel in apprehension . . . . a walking shadow, a poor player that struts and frets his hour upon the stage and then is heard no more. It is a tale, told by an idiot, full of sound and fury, signifying nothing.[1]

When we look at a simple object there is a temptation to think that we are the active agents, with our eyes acting in a manner somewhat analogous to a torch. In fact, the object is much more complex than it appears and between it and the eye, the eye is a passive receptor. Between the eye and the brain, the eye is the active agent and the brain largely passive. Between the termination of the optic nerve and the perception of the object there is substantial neural activity. These four stages – the nature of a material object, its transmission to the sensory organs, the onward transmission to the brain and finally the activity within the brain – will form the subject of the first four sections. The matter is largely regurgitated from the web, several textbooks and some works of popular science, and may be read more or less thoroughly; or may be largely skipped! What is important is some appreciation of scale and complexity.

As to the reality of the perceptions there are many different theories. At one extreme is the assumption that nothing apart from the material universe exists; at the other is the belief that the material universe is an illusion. We are caught between the extremes of what appears to be sound common sense (that the world is just as it appears to be) but which on closer examination seems to lead to absurd conclusions.

The undoubted scientific consensus is that there is no need to invoke the supernatural, yet for all the subtleties of much contemporary philosophy it is difficult to avoid the thought that materialism in the last resort reduces man to a very sophisticated automaton – very sophisticated but nevertheless an automaton subject in his every act to the laws of physics, with every action the result of earlier actions, right back to the Big Bang. This is not to say that they are predetermined: the reasoning behind this assertion is to be found in

---

[1] Hamlet 2.2; Macbeth 5.5

Part 2. The thought that all is predetermined borders on the absurd for in this case one would have no true freedom or responsibility. Thoughts themselves would have only a pragmatic value, validated by the inexorable march of evolution. For all the scientific consensus virtually nobody, not even most scientists, believes that this is the case.

A spiritual element in our constitution would obviate these problems but it would have to be shown not to be a deus (literally!) ex machina, an ad hoc device not supported by any evidence. There is indeed precious little evidence and what there is tends to be as reliable as that for UFO's. Worse than this, all religions are riddled with unwarrantable beliefs and superstitions.

That there is at least an underlying order in the universe is plain: otherwise no science would be possible.

There are a number of schools of thought about the reasons for this, which in essence can be reduced to three fundamental ones:

Current Quantum Theory or some refinement of it can or will explain it all.

The universe is the creation of a supernatural being who 'designed' matter, space and time.

It is just the way things are and it is vain to speculate further.

The first two have complimentary difficulties. If it is possible to construct a theory that can explain the eruption of something from nothing – not the emergence of matter from empty space, but rather space itself from nothing – it can still be asked where the theory itself comes from and its ultimate value. If the theory of evolution is the total truth, then from the simplest creature to the most complex, sensations are wholly explicable in terms of chemistry, and we are then in a closed system. The thought that we can understand ourselves has no more substance than the ability of any chemical reaction to be able to understand itself. It is a little like pulling oneself up by one's shoe laces. For understanding to have any authenticity requires some extra-material element. This statement is the core subject of Part 2.2, Clouds of Abstraction.

If understanding can be reduced to a series of chemical reactions, however complex, then it is all a chimera, possibly very satisfying but ultimately without substance: the flash of insight no more than a chance connection of brain cells. Of course, many insights might well be almost chance connections of brain cells but the conscious awareness of their import is something quite different.

As far as the second is concerned is it sufficient to say that God is his own reason and that is the end of that? Is it possible to say anything meaningful about something ineffable and inenarrable, quite outside our experience or power of comprehension? This is discussed in Part 4.2 Religion.

The third, the view of the common man (not a pejorative use of the term), might well be a considered judgement but surely cannot be an *a priori* principle, for who has the authority to place limits on what we might consider? From the all-is-illusory school of much eastern thought, through the tortuous progress of idealism of western philosophy from Plato through Berkeley to contemporary scientists such as Roger Penrose, no consensus has been reached and one might well be forgiven for coming to the reasoned conclusion that things are as they are, but we cannot come to it until we have dealt with the first two.

By trying to understand what we mean when we say we know something and so come to a judgement on these alternatives is the purpose of this essay.

PART 1

THE SCIENCE
Mysteries of the material world

There are more things in heaven and earth than are dreamt of in your
philosophy, Horatio.
What do you read, my Lord?   Words, words, words.

Knowledge of material objects starts with the physical perception of
them by means of our five senses.

1   The nature of the object perceived.

There has been a remarkable progress of scientific knowledge and
there exists an immense quantity of specialist literature, yet the clouds
of ignorance are so huge that it is difficult to know where to start.
Space and time, good and evil, matter and spirit, mind and
consciousness – there is no end but a start has to be made somewhere
so let it be the material world, with common-or-garden objects which
we perceive all the time. It could be a simple billiard ball, smooth,
rounded and monotone; it could equally be a tree-scape in early spring
with a thousand different shades of fresh green and the occasional
bare tree showing its intricate tracery against the sky.
By the material world is meant all the stuff out of which the universe
is composed, the stars and planets and everything to be found therein,
animal, vegetable and mineral. When we say we know some material
object what do we mean? We will start with our perception of it and
discuss later the connection between perception and knowledge.
Take the billiard ball. It is perceived as a small white smooth
odourless solid spherical object.
Until the nineteenth century the most popular substance was ivory.
The structure of ivory is far from monolithic: it consists of a matrix
of particles between five and twenty microns[2] in diameter which are
clearly visible in a good optical microscope.

---

[2] A micron is one-thousandth of a millimetre

12

The demand for ivory for this purpose was so large, and the pool of elephants so limited, that it became necessary to manufacture the balls and a very large prize[3] was offered for a suitable easily manufactured substance. Various plastics were tried and today the material most commonly used is a plastic resin[4]. This is very much more homogeneous than ivory but still, viewed under a powerful microscope the surface is pretty rough, and far from being ultra-smooth it will look more like a relief map of the alps. If the ball is cut in two the interior will be seen to be granular. Seen under an electron microscope the grains will become more apparent. These (and now we move into the realm of well-established theory) consist of complex chains of carbon, oxygen and hydrogen atoms.

An atom is the smallest indivisible unit of "solid" matter. By solid is meant matter which has mass. An object with mass needs a force to move it. Atoms are composed of nuclei consisting of comparatively massive particles, neutrons and protons, and a cloud of much smaller ones, electrons, encircling them at a distance.

An elemental substance is determined by the number of protons in the nuclei of its atoms. These vary from one, in the case of hydrogen, to eight in an oxygen atom, seventy-nine in gold to one hundred and eighteen in the synthetic element oganesson.

Compounds are formed from more or less stable combinations of atoms of different elements. Water, for example, is a very stable combination of two hydrogen atoms and one of oxygen. Whereas compounds can be split into their respective elements by chemical means, atoms cannot be similarly split.

Protons and neutrons are composed of very shadowy fundamental particles known as quarks; these can only be found in threes inside them and have no independent existence.

All commonly observable matter is composed of up quarks, down quarks and electrons. And to complicate the picture even further, electrons also exhibit wave-like behaviour which we cannot imagine and only understand as if in a cloud, darkly.

It is important to have some idea of the scale of the size of these particles if one is to have a glimmer of understanding of physics, of chemistry and of biology.

---

[3] Some $200,000 at current values (Wiki)
[4] Phenol formaldehyde resin

The finest spray atomiser produces a mist made up of liquid droplets of around twenty microns in diameter. To get some idea of the 'comparatively massive' size of an atom, a droplet of water measuring only one micron in diameter contains nearly one hundred thousand million atoms and there are living creatures considerably less than a cubic micrometre in volume yet having over one million base pairs in the genomes which make up their DNA[5].

As far as we know, photons and electrons are the two smallest irreducible units of the material universe. Partly because of their size and partly due to their nature we can only describe their behaviour in barely intelligible mathematical formulae, which lead us to believe that they act simultaneously both as particles and waves.

A photon is the ultimate non-divisible unit of energy. It has momentum but no mass and travels at the speed of light, which, over a range of frequencies, it is. It acts both as a wave and a particle, and can transfer its energy to the smallest, non-divisible, unit of mass, the electron. Every chemical reaction, every change in a living cell, is the result of the absorption or emission of a photon by an electron.

Photons vary in wavelength from one trillionth of a metre (1pm) in the case of gamma rays, to one hundred thousand kilometres in the form of very low frequency radio waves. The energy carried by a photon is inversely proportional to its wavelength and directly proportional to its frequency.

In wavelengths up to some billionths of a metre their energy is sufficient to affect atomic structure, causing, for example, a cancer. These are known as ionising wavelengths.

Visible light is in the millionths of metres, infra-red up to some millimetres, microwaves from under a metre to one hundred metres and then radio waves up to many kilometres.

The wavelength is inversely proportional to its energy in the case of a photon and to its mass and velocity in the case of an electron. What might not be quite so well known is that atoms too act as both, and molecules, and microscopic bodies, and macroscopic ones, in fact all material objects act both as particles and waves. The wavelength of a slowly moving electron is of the same order as the diameter of an

---

[5] For a brief explanation of DNA See 1.4, Evolution and the Living Cell.

atom. A neutron has some forty-thousand times an electron's mass and its wavelength is shorter by that many times, but still observable and used in some applications in some research. The wavelength is inversely proportional to the mass of a particle. Long before an object has reached one hundredth of a gram the wavelength has reduced by a factor of many billions of billions and is of no practical interest. Yet the billiard ball is in some sense truly a large particle and a wave of very short wavelength indeed.

The behaviour of particles of atomic size is the subject of Quantum Mechanics. The basics of the theory are difficult to grasp. Although its predictions are extremely accurate (perhaps a little like Ptolemy but to an extraordinary extent) it requires a fiendish mathematics and the interpretation, a full hundred years after its appearance, is still a matter of matter of great controversy. The more science advances, the more mysterious the universe appears to be.[6]

One of the prime experiments that led to the formulation of quantum theory requires only a simple apparatus: a light source which can be dimmed, an opaque sheet with two very narrow parallel slots and a photographic screen. When only one of the slots allows light through the light behaves as a stream of bullets would, and the photographic screen records a fairly sharp image of a line surrounded by a scatter of shots. If a jet of water rather than light was used a similar pattern would emerge. When both slits are open, the light behaves as a stream of water : passing through them would and an interference pattern appears on the screen: alternating dark and light bands.[7] If the screen had a fine grid of pressure sensitive cells and a stream of water was used an interference pattern of pressure difference would be seen: where the peaks of each wave coincide they are enhanced; when a peak from one is matched by a trough from the other they cancel each other. From this we are drawn to the conclusion that light behaves both as a stream of particles and as a wave.

Mysterious, but not nearly as mysterious as the following. While G I Taylor was still a graduate student at Trinity College, Cambridge, his

---

[6] CJK's oxymorons: Better science = more mystery
           More discipline = more freedom
[7] There is an excellent short You Tube illustration: *Diffraction: Why Does It Happen?* (Physics Explained for Beginners) by Michael de Podesta of the Institute of Physics.

supervisor, Professor Sir Joseph Thompson asked him to see what would happen to the diffraction pattern if the light was dimmed to the extent of only emitting single indivisible units, now known as photons, at a time. Taylor used a narrow slit in front of a gas flame and smoked screens to dim the light. This was shone onto a needle causing a diffraction pattern on a photographic backdrop. Similar diffraction patterns were observed however dim the light was: so dim in the end that Taylor was sure that "indivisible units"[8] were being emitted rather than a stream. This experiment took 2000 hours or very nearly twelve weeks.[9] The units were indeed single indivisible 'atoms' of light: photons. They were sufficiently separated in time so that one had already been photographed before the next passed the needle and could not have possibly have interfered with each other and yet interference was seen to have occurred.

Just as a lump of plasticine must have some shape, some form, at any instant so matter and energy cannot be formless. The underlying substance of each can only exist in one form or the other. In the case of matter, it must exist as a proton or neutron or electron – or in some other defined particle.

Energy must exist in the form of a photon of some defined wavelength. What is known about it is that matter can be converted into matter and vice versa in accordance with Einstein's famous equation $E = mc^2$, where E is energy, m mass and c the velocity of light. That is to say, the energy bound up in an object is the product of its mass and the square of the velocity of light.

In theory matter and energy are two fundamental forms of material existence; in practice the conversion of one to the other is only partially possible such as in the splitting of a uranium atom. Uranium 235 has a nucleus with 143 neutrons. It can be split into a Caesium atom with 85 neutrons and a Rubidium atom with 56, making a total of 141, giving a loss of two, which now exist in the form of heat.

This is the basic reaction which fuels current nuclear fission reactors. Unfortunately, the waste by-products of these reactors are highly radioactive and very difficult and costly to dispose safely. The much

---

[8] G I Taylor *Interference Fringes with Feeble Light. 1909*

[9] In his excellent book, *The Pleasures of Counting,* Professor T. W Körner cites Taylor's biographer claiming that he chose the project to allow him time for a couple of months cruising on his yacht. On the face of it this seems unlikely.

safer nuclear fusion where hydrogen atoms are fused to form helium with the loss of some mass is proving very much more difficult to achieve.

To give some idea of the total quantity of energy which is contained by a given mass, one kilogram has twenty-five billion kilowatt hours of energy locked up in it – nearly enough to power a city of one million inhabitants for a month. If only it could be harnessed and released safely!

The billiard ball, composed of billions of billions of these extraordinary objects, is far from being a simple undifferentiated white spherical object: it is largely empty space with billions of little whirlpools of activity acting and interacting on each other.

Well, let us just pot it. It has to move through space and will do this at a certain velocity – velocity rather than speed because velocity is speed in a particular direction. Space and time. What is known about their nature?

2   From perceived to perceiver.

2.1   Modes of transmission

Objects manifest themselves to living creatures in several diverse ways.

Take a hot potato, rather than a billiard ball in this instance, a baked potato, still in the oven. When the oven is opened there is an instantly recognisable odour. The potato is visible, partly by daylight in the kitchen and partly by the oven light. It looks, if anything, overcooked. Take it out with your bare hands. Say 'Oh ouch' (or words to that effect) and drop it onto the tiled floor where it lands with a dull thud. Pick it up using a tea towel, put it on a plate, cut it open, add a little butter, sprinkle it with salt and pepper, and eat it. Delicious.

Light from the potato impinges on the eye. In general, light might come directly from the object, as in the case of it being red or white hot. More commonly, in the daytime it is a reflection of sunlight or at night a reflection of artificial light. Some of the light rays (photons) which hit the object are absorbed by electrons on its surface and then emitted and directed to the eye. The hotter an object is the greater is its radiated energy in the form of heat, and this will reach an observer across small or large distances (the sun is ninety-three million miles from us) as well as the exchange of photons across touching surfaces.

Sound is the result of air-borne pressure waves travelling from emitter to ear. Not only does this need the presence of a gas (usually air) between the two (noise cannot travel through a vacuum), but a fairly still gas. It is possible to speak quite normally to a passenger in a fast-moving car as long as the windows are closed. If they are opened and there is a rush of incoming air, it is disturbed, and speech becomes impossible. As an aside, the thorny question of whether a falling tree in an animal-free forest makes a noise can in part be considered here. It will undoubtedly produce a pressure pulse in the surrounding air which will propagate through the forest. Whether this is a noise will be discussed later.

Smell and taste are both caused by chemical reactions which activate cells in nose and on tongue. Odours consist of airborne volatilised chemical compounds, usually, but not always, organic – that is, compounds consisting of carbon atoms bound to each other or to other atoms.

Taste is very similar and although it can occur at a distance it is usually by contact between object and tongue. Contact, of course is not as simple as might be thought. At a molecular or atomic level there are clouds of electrons in close propinquity and a complex interaction. Touch utilises cells in the skin which react to pressure and others which are sensitive to heat or cold and needs direct contact. There are instances where the nerve cells which are activated by direct pressure can be activated at a distance. Marvell describes one:

> Now therefore while the youthful hue
> Sits on thy skin like morning dew
> And while thy willing soul transpires
> At every pore with instant fires. . . .

In this example sight, sound and even thought all play their part.

2.2   Fundamentals of space and time

Movement requires space and time. Time has puzzled man for all time; space less so, although perhaps it ought to have done.

Someone (it might have been Lamb) said that above all things time puzzled him most, but as he never thought about it, it could be said to puzzle him least! Augustine says[10] that when asked what God did in the aeons before he created time, he answered that someone said that He was preparing a hell for those who asked such questions but that he Augustine did not approve of such frivolity in what was an important question. He went on to say, in Book 11 of his Confessions in his great enquiry into the nature of time, that truly there was no time before time began. The same might be said of space. Starting with both on a grand scale, observe the rising of the sun each morning. The increasingly complicated theories to 'explain' the movement of the planets culminated in Ptolemy's geocentric series of epicycles within epicycles and were developed in the Second Century AD following Greek attempts a couple of hundred years earlier to give them a mathematical basis. They forecast planetary movements with sufficient accuracy to enable the forecasting of eclipses and the like and provide necessary astrological information. The understanding of the Ptolemaic mathematical model was a measure of a person's knowledge and intelligence for the next several hundred years.

The whole edifice was total nonsense.

---

[10] Confessions.

If not total nonsense it was a prime example of making a theory fit the observed facts, not by altering the theory but by ad hoc additions to it: Planetary motion must be perfect. Perfect motion, it was assumed, is circular. Ergo. Unfortunately, this does not accord with observation of their orbits. Let them be epicyclical; that is, making smaller circles whose centres are travelling round a larger one. Add further epicycles and it is just a matter of the number, size and rotation of them to achieve paths which appear fairly simple to an earthly observer, and to be able to be plotted accurately in accordance with their actual paths in the sky.

Copernicus had the bright idea of putting the sun rather than the earth at the centre but his assumption that all the orbits were circular also led to the necessity of substantial tweaking.

At the height of the Renaissance, towards the end of the fifteenth century, men like Leonardo began to observe living creatures and to correct notions unchanged since classical times. It seems extraordinary to us that from classical times until the fifteenth century, some seventeen hundred years, men relied on the authority of a number of Greek philosophers, of whom Aristotle is the foremost, and it was not until the late sixteenth that observation of the patterns of behaviour began to be codified and described in terms of laws.

There is a story of a number of fourteenth century philosophers (the Scholastics) arguing at length about the number of teeth in a horse's mouth until one of them had the bright idea of going outside, finding one and counting them!

It was not until Kepler formulated his three laws,[11] followed by Galileo, and then Newton (who was born the year after Galileo died) who realised that if there was an attractive force between bodies, proportional to their masses and inversely proportional to the square of the distance between them, the sun and the planets would follow paths in accordance with a very simple formula. From Copernicus in

---

[11] 1 The orbit of a planet is an <u>ellipse</u> with the Sun at one of the two <u>foci</u>.

2 A line segment joining a planet and the Sun sweeps out equal areas during equal intervals of time.

3 The square of a planet's <u>orbital period</u> is proportional to the cube of the length of the <u>semi-major axis</u> of its orbit.

the early sixteenth century to Newton in the late eighteenth the development of an accurate heliocentric model based on a simple set of rules took over a hundred and fifty years.

Newton proposed a theory of gravitation and three simple basic laws. These state that objects in motion would continue in straight lines unless acted upon by a force, that their motion is such that their acceleration is proportional to the force acting on them divided by their mass, and thirdly that every action has an equal and opposite reaction. These four notions received acceptance as a complete and universal explanation of the behaviour of the all material bodies, although the nature of gravity and its ability to act at a distance through space remained (and still remains) a mystery. The idea of a clockwork universe arose: once created it would proceed along set cause-and-effect lines. The behaviour of all matter could in theory be calculated. Space and time were given absolutes. This led to all sorts of philosophical problems which tied men like Kant in knots, but after Newton came Einstein:

> *Nature and nature's laws lay hid in night*
> *God said, 'Let Newton be." and all was light.*[12]

> *It did not last. The devil shouting, "Ho.*
> *Let Einstein be." restored the status quo.*

In the late nineteenth century Newtonian physics received two quite separate death blows to their being the ultimate explanation of material behaviour, although extraordinarily accurate for small scale macroscopic situations. First it became increasingly apparent that Newtonian physics did not account for the behaviour of objects of atomic size and early in the last century the statistical nature of quantum physics emerged and the idea of universal determinism became untenable.

Of course, in practice, our knowledge of all physical actions and reactions has been statistical. Anyone who has carried out an experiment in a laboratory will be aware of this. However carefully the parameters are set to ensure non-interference from any external source and however accurate the measuring instruments, some variation will be found and a mean result calculated. For example, if the time taken for a billiard ball to drop a given distance in a vacuum

---

[12] Alexander Pope and J C Squire

in a glass tube, the answer would be of the form so many seconds plus or minus some percentage even if this is only a few thousandths. This fact was clearly articulated by Richard Feynman[13]: *It is not therefore fair to say that from the apparent freedom and indeterminacy of the human mind, we should have realized that classical "deterministic" physics could never hope to understand it, and to welcome quantum mechanics as a release from a "completely mechanistic" universe. For already in classical mechanics there was an indeterminability from a practical point of view.*

Secondly it slowly came to be realised that when Einstein produced a theory that neither time nor space were absolute, and that both were inextricably bound up in each other, his logic could not be denied and the theories of Special Relativity dealing with uniform motion and several years later General Relativity, with accelerations, gained general acceptance.

Newtonian physics is readily comprehensible and at least the outlines known by most educated people. The Special Theory of Relativity is more difficult but the famous conclusion, $E = mc^2$ is universally known even if a majority of educated people would be hard put to it to give its meaning, let alone its implications.[14] The General Theory is very much more difficult, and very few of us have any grasp of it: the mathematics involved are not for the ordinary citizen or even the first year student taking a maths degree.

One of the extraordinary properties of light is the speed at which it travels. Not only that in our experience of speed it is extremely fast at three hundred thousand kilometres (one hundred and eighty-six thousand miles) per second (per second!) but the unimaginable fact that this speed is absolute. For everything of which we have experience speed is relative. Being in a car and held up on a motorway by an articulated truck travelling at sixty-one miles an hour overtaking another going at sixty can be irritating: the one barely crawls past the other. If these two vehicles pass someone standing on the hard shoulder they thunder past at a frightening speed.

If a fighter plane fires at another which is passing it at a greater relative velocity than that of the bullets, they will never reach it. Even if the plane is disappearing at a speed of a hundred and fifty thousand miles a second., a ray of light from a laser will still reach it at a speed

---

[13] Lectures on Physics 1.39

[14] At least in England – the educated elite in many other countries are not so ignorant.

of a hundred and eighty-six thousand miles a second. The speed of light is an absolute. (The question of what would be the case if the disappearing plane is travelling at the speed of light does not arise: it is not even theoretically possible.) There are two unimaginable consequences of this fact. By unimaginable in this case is meant that we, humans, can form no picture of it, not that we cannot accept it as a proven experimental fact.

The consequences are time dilation and and length contraction.

If a watch initially at rest with another is accelerated in a space rocket and observed and measured from earth, it will record time at a slower rate – as will the heartbeat of the astronaut wearing it, and everything else in the rocket, the frequencies of the sub-atomic wave-particles of which everything is composed, including the atoms and electrons of the rocket itself. This results in a person moving very fast with respect to those on earth ageing more slowly than those on earth, although for the effect to be noticeable the speed has to be approaching that of light. At a speed of ten thousand miles per hour the difference is negligible: the moving watch will record less than one-tenth of a second after a month of travelling; at half the velocity of light, ninety-three thousand miles per second, thousands of times faster than a high velocity bullet, the watch will record forty-five seconds to each earth minute. Were it possible to for you to travel at nine tenths of the speed of light for twenty years, when you returned to earth you would have only aged seven years and at ninety-nine hundredths, something over two years.[15]

Consider a short fable, a modern version of Dorian Gray. Phileas Finn updated.

> *Six members of the Reform Club were playing cards and having a fierce argument about a space craft which Phileas had recently bought and which he claimed could travel at 175,000 miles per second. To prove it he would travel round the newly discovered dead star Proximus Ω, nearly one light year distant, in not more than three hundred and sixty-five days.*

---

[15] The dilation factor is $\sqrt{1 - \frac{v^2}{c^2}}$ where v is the speed of the moving object and c that of light.

"I'd like to see you do it in not more than one year"

"It depends on you. Shall we go?"

"Heaven preserve me! But I would wager four thousand pounds that such a journey is impossible."

"Quite possible, on the contrary," returned Mr. Fogg.

"Well, make it, then!"

"The journey round Proximus Ω in one year?"

"Yes."

"I should like nothing better."

"When?"

"At once. Only I warn you that I shall do it at your expense."

"It's absurd!" cried Dorian, who was beginning to be annoyed at the persistency of his friend. "Come, let's go on with the game."

"Deal over again, then," said Phileas Fogg. "There's a false deal."

Dorian took up the pack with a feverish hand; then suddenly put them down again.

"Well, Mr. Fogg," said he, "it shall be so: I will wager the four thousand on it."

"Calm yourself, my dear Dorian," said Fallentin. "It's only a joke."

"When I say I'll wager," returned Dorian, "I mean it."

"All right," said Mr. Fogg; and, turning to the others, he continued: "I have a deposit of twenty thousand at Baring's which I will willingly risk upon it."

"Twenty thousand pounds!" cried Sullivan. "Twenty thousand pounds, which you would lose because it cannot be done!"

"Cannot be done is not the case," quietly replied Phileas Fogg.

"But, Mr. Fogg, one year in which the journey can be made depends on a faultless journey at a speed which no vehicle has yet achieved."

"My vehicle will be faultless and achieve it."

"You are joking."

"A true Englishman doesn't joke when he is talking about so serious a thing as a wager," replied Phileas Fogg, solemnly. "I will bet twenty thousand pounds against anyone who wishes, that I will make the return journey in one year or less; In three hundred and sixty-five days or eight thousand seven hundred and sixty hours or five hundred and twenty five thousand and six hundred minutes Do you accept?"

"We accept," replied Messrs. Dorian, Fallentin, Sullivan, Flanagan, and Ralph, after consulting each other.

"Good," said Mr. Fogg, "I have already provisioned my vehicle with supplies for one year and will leave this evening."

"This very evening?" asked Stuart.

"This very evening," returned Phileas Fogg. "As today is Wednesday, the 2nd of October, I shall be due in London in this very room of the Reform Club, on Saturday, the 2nd of October next, at a quarter before nine p.m.; or else the twenty thousand pounds, now deposited in my name at Baring's, will belong to you, in fact and in right, gentlemen. Here is a cheque for the amount."

A memorandum of the wager was at once drawn up and signed by the six parties, during which Phileas Fogg preserved a stoical composure.

Phileas set off out of the club at a brisk trot.

Nothing was heard of him for the next twelve months and Dorian returned triumphantly to the club to demand the prize from the stakeholder. Three months later the triumph turned to anxiety as there was still no news of Phileas, who must by now have run out of food and water. Each year for the following four a commemoration dinner was held on the anniversary and a toast drunk at a quarter before nine in the evening. Dorian was consumed with guilt at goading his great friend Phileas into such a rash wager, which had resulted in his death. He brooded incessantly and grew thin and haggard.

As the travellers sat down to begin the commemorative dinner, Phileas strolled in, looking pale but fit and well fed.

'I actually landed more than a week ago,' he said, *'but laid low until now! To prove I have been round Proximus Ω in less than a year, here is my sealed watch and tracker.'*
While one year had passed in the spacecraft, five had elapsed on earth.

Who won the wager?

The seeming paradox of which person is really travelling is not really a paradox at all: only one of them must have experienced forces due to acceleration so they are in different situations.

The obvious question arises, 'What is the medium carrying the waves if not space?' Here we come up against the limitations of language. Things can only be described in terms of other things, and this can be by visual comparison – 'That is a blue tit' – or by verbal description using a known object or objects – 'Ratatouille is a dish made from aubergines, peppers and tomatoes.'

When it comes to space, we have nothing to compare it to and no words to describe its essence. Saying that it is "Extension" or that it is the medium in which electromagnetic waves and gravity are carried does not take us much further forward but then it might be argued that this is true when it comes to the ultimate understanding of any substance.

There is a great deal we can theorise about its nature, in mathematical terms, as set out in the as yet unbridgeable theories of relativity and quantum mechanics and perhaps this is the most that is available to human intelligence.

Que sara sara is not an expression of determinism but rather a statement that there is sense in the notion that just as the past and the fleeting present exist, so does the future. What will be will be: how it will get there is another matter.

The notion that the entire universe, its past and present and future have an equal reality and that we, conscious creatures, only become aware of a little part of it, little by little, is known as the block universe.

The Special Theory of Relativity was the first to bring the notion of the relativity of time to a widespread audience. The idea that there is no universal 'now' has led to the development of theories of the status of the future in which the implications of the common sense 'que sara sara' are more closely examined. The future is regarded to be as real

26

as the past and the theory of a block universe given a fairly clear articulation: the entire universe from its beginning to its end is a four-dimensional space-time construct. It is very controversial with many philosophers and scientists dismissing it as being without any evidence.

The idea of a block universe is difficult to grasp but there is a fairly straightforward model which can help.

Imagine a book made up of eight hundred copies of an ordnance survey map of London so that each page represents one year from 1500 to 2300. Now on each page plot such events as might be found in a history book in their correct locations. The coronation of Henry 8th in Westminster Abbey on page 9, the execution of Thomas More at Tyburn on page 35 and the Fire of London further to the east on page 160 and so on until the present day. What will happen in the next two hundred is not yet known, but happen it will and in the block universe has as real an existence as the fleeting present and the unrecapturable past. We cannot imagine a block universe because rather than a book made up of many flat layers, a four-dimensional model is needed, with three of space and one of time.

It might be objected that a true model would need to have time intervals of considerably less than a year, and the question arises as to whether time is infinitely divisible or like energy is quantised. There is no consensus on this. A chronon is a proposed quantum unit, with a value of less than one billionth of one billionth of a second.

My 'now' is a subjective experience, an infinitesimal or possibly quantum period of time. In so far as it is subjective, it is different from your 'now'. In so far as we cannot inhabit the same space and have different motions, my 'now' might be your past and your past my future. You are aware of my future just as I am aware of your past. Time is both real and illusory!

It is tempting to dismiss non-scientific beliefs, by which is meant not supported by experimental evidence, as not being worth serious consideration. The view that time is illusory has been held by eastern religious teachers and by philosophers of many nationalities for aeons; it has been professed by mystics of all religions and by Einstein!

In Hinduism the past, the present and the future coexist in God simultaneously.
(This is also the explicit teaching of Augustine, later echoed by Thomas Aquinas)

Says the Yoga Vashista, *The world is nothing but a mere vibration of consciousness in space. It seems to exist even as a goblin seems to exist in the eyes of the ignorant. All this is but Maya: for here there is no contradiction between the infinite consciousness and the apparent existence of the universe. It is like the marvellous dream of a person who is awake.*[16]

That space and time are intimately connected is commonplace: 'Meet me on the corner of Broadway and West 42nd Street' is as useless without specifying a time as 'Meet me at half past ten' without naming a place. Even without the complications of time, the nature of space at an instant of time is difficult to understand, even hazily. Just as time had a beginning and there was no time before time began so space had a beginning and there was no space before space came into being.

In ordinary language, space is not nothing. If it was there would be nothing separating us from the moon: we would be touching[17]. Let nothing come between you and me. This is not just a verbal trick. Space is not a void. It is finite in extent and composed of a number of electromagnetic fields and the closely analogous gravitational fields of which, to say the least, current understanding is far from complete. A simple example of a field is the temperature at every point in a draughty room, varying from a comfortable 22° C (72° F) near the fire to freezing close to the windows and warmer at ceiling level than floor. It is, of course, the air that is cold or warm and the heat reaches the particles of air by conduction, convection or radiation, each with their own fields. A field is the extent and pattern in which some physical presence is spread out in space. A colossal number of electromagnetic fields pervade all space; together with gravity they are space. The fields carry energy in the form of photons.

This substance, space, which seems at first sight to be just emptiness, turns out to be an almost (for most of us a totally) unintelligible complexity or perhaps more likely, a nearly unintelligible simplicity! 'Space is the medium in which electromagnetic and gravitational waves carry energy,' or perhaps better, 'Gravitational waves are space-time in which electromagnetic waves travel.' Energy is often

---

[16] Yoga Vashista, Part 3, On Creation, The Story of Lila, Translation by Swami Venkatesananda, State University of New York Press, Page 86

[17] Descartes beat the present author to this idea by some 300 years.

defined as that which has the potential for doing work: for moving an object against gravity, for heating an object, for making one shine or emit a noise. A little like time, everyone knows what energy is but the understanding of its nature is very difficult, if not impossible as we have no sensory apparatus to perceive it. Of course, we can see light, feel heat and blows, hear sounds, smell roses and taste curry, all of which are forms of energy.

We are moving out of the realm of the structure of material objects, its enormous complexity and for all the research and discovery over the last two or three centuries, its expanding mysteries.

We know that much of what our ancestors believed to be the structure of the universe, such as the elements being earth, air, fire and water, was factually wrong; we think we know that that the consensus on current scientific knowledge is factually right; we know that there is still an unfathomable depth ignorance in our knowledge.

Knowing a material object means that we have in our memories a variety of sense impressions and facts or beliefs obtained from many different sources.

What energy is in itself is not a subject for this chapter, any more than what matter is without any form.[18] As has been said, just as a lump of plasticine must have some shape at any instant so matter and energy cannot be formless. The underlying substance can only exist in one form or the other.

What then is meant when we say we know something?

It is true that the word *know* is commonly used in the sense of knowing the existence of something: *You know the oak tree on the green?* Or, *You know the Royal Oak?* Here the word *know* can mean anything from a hazy recollection to a detailed knowledge of the bar and the appearance of the landlord.

When we say we know some material object it is that we have some image of it embedded in our memory.

We have examined the manner in which these images travel from any material object to the sense organs of any living organism. Now let us consider the passage from them to the brain.

---

[18] Aquinas' 'Prima materia' expanded to include matter and energy. Cf Summa Theologica 1 LXVII 1

3   Organs of sense: from skin to brain

Our five sense organs, eyes, ears, tongue, nose and skin have resulted as the products of evolution. Its course has been extensively documented but some aspects of it are the subject of fierce controversy, particularly in the classification of species. This controversy does not affect the main theory at all and although a sizable proportion of the world's population, even in 'advanced' countries such as the USA do not believe in evolution at all, preferring to take creation accounts such as that in Genesis as being literally true, evolution will be accepted as factual in this essay. In those who deny it, not only is the enormous amount of evidence from many different disciplines ignored but sometimes bizarre theories are put forward, such as that of the Nineteenth Century naturalist, Philip Henry Gosse who claimed that God had created the fossil evidence to lead the unfaithful astray.[19]

3.1   Evolution and the Living Cell

The current scientific consensus is that almost fourteen billion years ago our universe erupted into nothingness, into a void devoid of time or space. Initially the very high temperature and density universe expanded rapidly.

For the first ten billion years the evolution was in the form of the establishment of the stars and their planets and of the forms of the latter. Current belief is that life on Earth did not emerge until some four billion years ago. How it emerged is still hardly a matter for serious theories based on credible models. This is not to say that it is not a subject for serious research but rather that any evidence is almost completely absent. What element an atom is, is determined by the number of protons and equal number of electrons in its nucleus. Depending on the number of electrons in the outer layer of the cloud surrounding the nucleus elements are more or less likely to combine to form molecules and the likelihood of random collisions of atoms forming molecules is fairly high, and of molecules combining to form ever larger stable compounds is easy to imagine. The likelihood of random collisions of molecules joining to form DNA and RNA is

---

[19] Omphalos, 1857

30

exceedingly remote. Yet, however unlikely any chemical reaction might be, given the right conditions and sufficient numbers and sufficient time it will happen. Ten thousand million years and very considerably more than this number of molecules confined and colliding on planets such as Earth proved to be sufficient! It is possible that the very nature of matter favoured the evolution of life just as it favoured the formation of molecules, making both inevitable. The diameter of a water droplet in a very fine mist (dry fog) is of the order of one hundredth of a millimetre, the same as a very fine speck of dust. It will contain an almost unbelievable number of water (H_2O) molecules, each composed of two hydrogen and one oxygen atom bound together: in excess of a thousand million. A thousand million atoms in a mist-droplet of water or a speck of dust.

At first, atoms were formed, some of which collided to form molecules; galaxies, stars and planetary systems came into being and in the material which covered the surfaces of some of them, the primordial soup, some molecules joined to form mega-molecules and a mega-molecule (or a few, or many) combined to make the miracle which is DNA, deoxyribonucleic acid. This is a two-stranded ribbon molecule made up of pairs of matched nucleotides, organic molecules, adenine, cytosine, thymine and guanine (A, C, T or G) such that A is always linked to T and C to G. When a cell divides the strands separate and each picks up free nucleotides to form an identical double strand. One DNA strand can contain hundreds of millions of base pairs the order of which in turn determines the nature of the creature: a grasshopper or a hippopotamus.

The simplest known life forms, such as unicellular bacteria range from one micron in size (one thousandth of the size of a water droplet in a fine mist) to over one hundred microns. All are extraordinarily complex. Each has a number of separate components: Cytoplasm, Ribosomes, DNA, Plasma membranes, Cell membranes, Cilia and a 'tail' known as a Flagellum, each with its own function. Each bacterium has in the order of thousands of billions of atoms. It can reproduce, feed, grow, live its useful or harmful lives (seen from our perspective), age and die. Bacteria exist in huge numbers all over the globe, on earth, under the sea and in the air. They do not have a central nervous system but are able to react to external factors and to this extent can be deemed to have some elementary consciousness.

There is a clear progression in terms of increasing ability to react (broadly speaking, consciousness) from unicellular animals through fishes and birds to mammals and humans is a part of this process.

It would be tedious to move along the evolutionary course phylum by phylum and trace the increase in consciousness so only two or three will be briefly mentioned. With the increase in consciousness came pleasure and pain, though at which stage in evolution is not so clear.. Almost no one believes that earthworms, Annelida, feel any pain when cut in two by a spade; opinion is divided when it comes to fish, Chordata; few would deny that a dog, again Chordata, can suffer and be cruelly treated.

It is thought that in theory it would be possible to trace the genetic evolution of any given earthworm, fish or dog, gene by gene right back to the Big Bang. By studying their DNA we can only do this taking leaps of millennia, but the general pattern is clear. Increases in brain size and nervous systems lead to an increase in sensitivity towards external stimuli: increase in consciousness. At some stage in this process we, educated western mankind, have decided that pain can be felt, and before this (before, not necessarily in a temporal sense) it cannot. It is difficult to see any logic in lawmakers deciding that one member of the phylum Chordata, fish feel no pain when dragged out of the ocean and die flapping in agony on a fishing vessel's deck, whereas another member of the same phylum, a donkey, is portrayed in television advertisements suffering horribly in their abandoned old age after a life of heartless labour. The thought that alteration to a single gene, results in a creature which could feel pain and whose parents could not, is clearly absurd. The logical conclusion of this would seem to be that all the animal kingdom can feel pain or none can.

Without trying to decide which of these is true it is well to pause and look at the notion of cruelty. Only humans (or gods: *As wanton boys to flies are we to the gods; they kill us for their sport*[20]) can be cruel in its proper sense. A fox killing but not eating hens, a thrush dragging an earthworm out of the soil or a cat playing with a mouse are not cruel; a man beating a horse might be: it depends on his state of mind. The use of a whip in horse racing is now limited by law but not prohibited altogether and the only distress shown by the winning whipped horse appears to be tiredness. While being led into the

---

[20] Gloucester, in King Lear 1V 1

32

winners' enclosure it tosses its head, giving every appearance of pride.

From birth, the body is bombarded with sights and sounds, buffeted by blows, subjected to tastes and smells; some of these affect the embryo as do some effects of radiation. They arrive at the body in various forms, sub-atomic particles in the case of light, chemical molecules in taste and smell, pressure in touch and hearing. All are transformed into electrical pulses by vast numbers of specialised living cells in our skin and eyes and ears. The pulses travel along various nerves to the brain where they lay down memories and sometimes produce conscious impressions: we smell a rose. Sometimes we immediately recognise it as an *Ena Harkness* but more commonly just as a rose and might (or more probably might not) wonder what sort it is. Whichever alternative it is, memory, conscious or unconscious, is needed.

To try to understand how an object reaches our conscious perception as a beautifully formed red, fragrant flower with closely furled petals and a many-thorned stem will be the subject of this chapter.

Pluck it, taking care not to get spiked. Observe it closely, feel it, cradle it in your hands, smell it; do not take a bite.

Eyes receive reflected light coming off it. The eyes are virtually passive in respect of the flower: apart from a little reflected light coming off one's eyes, which might add to its brightness, the appearance has not been generated by the eyes at all. Reflected sunlight from the flower does affect the eyes. They are active in respect of our sensations. Nerve cells at the back of the retina react to the light and send electric pulses along the optic nerve to the brain. The brain is largely passive with respect to this transmission although neural responses in the optic nerve might be inhibited or triggered; it is active once the impulses have reached it. When the brain receives the impulses some of its billions of cells are activated and these in turn are transmitted to others through an unimaginably vast complex of interconnections. Memories are stirred, pulses travel back down various nerves, and muscles are stimulated: the eyes move and focus, hands rotate, heads duck. The broad paths of some of the electric pulses can be followed using medical scanning devices but the precise process undergone before a rose is seen (consciously or unconsciously) is still largely unknown. We do know that it is possible to see objects and not to be aware that we are seeing them. A good example, familiar to anyone who drives a lot, is highway hypnosis. Considerable distances can be driven while the driver is away in a daydream. When consciousness returns, sixty would-have-been tedious miles might have been travelled with no recollection at all of that part of the journey, yet other traffic, traffic lights and road signs have all been seen and the appropriate actions taken. Sleepwalking, somnambulism, is another common example with reports of even driving a car whilst asleep.

No one as yet has the slightest idea how the neural activity enters the sphere of conscious awareness. For all the advances in knowledge over the last three hundred years the nexus between neural activity and self-conscious awareness is still a complete mystery.

Each of the two optic nerves consist of over a million fibres, leading with no internal connections directly to the brain.

The same is not true of sensory nerves responsible for conveying sensations of touch or heat. If you touch the rose a similar process occurs via pressure and heat sensitive nerve cells in the skin. These nerves are composed of short lengths connected by synapses, numbered in hundreds and even in some cases, thousands, which relay the electrical pulses to the brain. Smell the rose and olfactory cells in the nose transmit pulses to the brain enabling us to smell it. Snap the stalk and a slight noise will affect the ear drums causing yet other neural responses in their cells. Taste can sometimes be useful in identifying objects but obviously needs to be used with extreme caution.

To begin to appreciate the complexity of any sensation some idea of scale might be helpful: the comparative minuteness of the elements of matter and the enormous numbers involved. A water droplet 1mm across will contain nearly twenty million millions (20,000,000,000,000) water molecules each composed of two hydrogen and one oxygen atoms.

The human eye is an approximate sphere of a little under 25mm diameter. Each one has a lens, the pupil, the small dark circle in the middle of the eye which focuses light on to the retina. In front of the pupil is the blue, brown, hazel, grey or green iris, which can open and close rather like a camera shutter to allow more or less light to enter. The iris is surrounded by the sclera, the white part of the eye which is painfully sensitive to touch. The retina, at the rear of the eyeball has around one hundred million light sensitive cells (rods and cones), which convert incoming light to electric pulses. These pulses travel along the optic nerve to the brain, and reciprocal pulses operate the various muscles which control movements in the different parts of the eye.

The most powerful current digital cameras such as the Hasselblad H6D have a similar number of photo-receptors to the human retina but unlike the eye these are not living, constantly self-renewing cells and do not in their processors have remotely the extraordinary web of interconnections. The H6D can take around one and a half frames a second; the eye processes incoming data in the low hundreds of 'frames' per second and transmits these to the brain. The brain itself is the equivalent of a very low speed computer of up to two hundred operations per second as opposed to a fast computer at four billion. However, the brain acts as a multi-core processer with the order of a hundred billion cores!

The human tongue has far fewer taste buds (gustatory cells, chemoreceptors) than the eye has photoreceptors: up to ten thousand, which transmit electrical pulses along cranial nerves to the brain. Taste buds are believed to be renewed at fortnightly intervals. Each nostril too has on average some eight million olfactory chemoreceptors; the clitoris and penis each have a similar number of touch sensitive chemoreceptors which transmit impulses to the brain. Some internal organs transmit sense data to the brain so that aches and pains are felt but for the most part this is not the case.

The nature of the objects sensed has been discussed in 1.1. To recap, it will suffice to say that there is no simple or straightforward description of any material object. All material objects within the scope of our senses are composed of billions of tiny particles (tiny, when compared to a person): molecules, atoms, electrons and other sub-atomic particles. Yet, what you see and feel and smell is really there. Is there any sense in asking whether it really is as you see it? After all, there are minor differences if you look at with one eye closed and then with the other eye closed; in some forms of synaesthesia the rose can be heard as well as seen. It might be argued that synaesthesia is a disease, and sufferers experience an hallucination rather than reality, but if the great majority of the population had the genetic disorder that causes synaesthesia it would not be called a disease and the genetic sequence not treated as a disorder. To hear roses would be the norm.

Our genetic coding is the result of millions of generations of parents passing their genes down to their offspring. Generally, the genetic transfer is accurate; occasionally there is a malfunction, due to a variety of reasons. Cosmic rays may dislodge a gene pairing, or a slight malfunctioning of the copying mechanism has occurred, and a minute variation, a genetic mutation, takes place. This malfunction is random, and the offspring will have some slightly different characteristic (it might be eye colour, or a preference for stewed spinach). It might have some advantage in the struggle to find food or a mate or avoid an attack and it is the driver of evolution. After dozens of generations new species emerge (groups of animals which by and large cannot interbreed). Over time, sometimes only one generation, sometimes many thousands, some species survive while other become extinct, again due to a variety of reasons. It might be the slow erosion by disease, depredation by predators or cataclysmic disruption by the impact of a large meteor. If, over the last few million years a human species with synaesthesia had evolved into a separate species and our current species had become extinct, then all visible objects would be noisy.

The course of evolution is very uneven and the fact that some species survive, evolve and thrive, has many facets. The conferring some small advantage in the struggle to find sufficient food or to be more able to fight off a predator – the survival of the fittest – is the commonest explanation but not the only one unless 'fittest' is tautological: if a species has survived it must by definition be fitter than species which have become extinct! A species might become extinct, as was possibly the case with dinosaurs, because of a cataclysmic event such as the shock caused by the impact of a huge meteorite or a catastrophic viral attack. Genetic variation might protect against the latter, but certainly not the former.

Species become extinct both as part of the natural evolutionary process, due to a variety of reasons, and because of human activity, with the latter as the current primary contributor. Prior to this the background or normal extinction rate was measured in the low millions of years per species.[21] This has been dramatically shortened by human activity.

It can be readily understood that feeling aches and pains help protect animals from harming themselves and from letting untreated illnesses

---

[21] Wiki, Background extinction rate.

get worse. Not only humans use medicines, although only humans manufacture them. In this aspect evolution could be thought to have performed poorly in respect of the many cancers which grow without giving any cause for alarm and which could have been easily cured had early warning been given. On the other hand, the ineluctable growth of the human race is testament to its success. A huge nuclear conflagration, against which no animals are proof, would be an example of a non-genetic reason for the extinction of a species, but since only humans could have invented and built the bombs, perhaps evolution could be seen as responsible.

There are two main causes of sensations: interaction of cells within our skins with physical pressures (touch and sound) or with chemical particles (smell and taste), and interaction of cells with some part of the vast electromagnetism which pervades the universe (sight).

Electromagnetic waves consist of small individual units of energy, photons, which have no mass, and which individually travel through space simultaneously as both particles and waves at a speed of three hundred million metres per second (the speed of light). The quantity of energy of each photon is inversely proportional to its wavelength. The wavelengths vary from approximately one billionth of a millimetre in the case of cosmic rays to radio waves of tens of thousands of metres, with visible light somewhere in the centre.[22]

Gravity too has a wave structure but is still even more mysterious than are photons. It too probably consists of indivisible units, quanta; it travels at the speed of light and has its own form of energy, but the science of quantum gravity is a matter of some controversy with still a great deal to be discovered.

Man's sensory apparatus cannot detect the majority of the vast array of electromagnetic waves which permeate all space. In fact, apart from the visible or near visible spectrum none are available to our senses. From ultra-shortwaves such as gamma rays (which can cause mutations in genes and initiate cancers) to longwave radio transmissions (which have no known effect on us) we are totally unaware, although some time after exposure we can become aware of the effects of ultra-violet transmissions which result in sunburn and in skin cancers.

---

[22] The wavelengths of visible light are between 400 and 700 nanometres A nanometre is one-millionth of a millimetre.

We only become aware of radio waves when they are intercepted by an appropriate aerial and transformed by a radio into sound waves. Sound consists of pressure waves in the air which cause ear drums to vibrate. The vibrations are transmitted to the small bones in the middle ear, amplified and passed to the nerve cells in the inner ear whence they are converted to electrical impulses carried to the brain and we hear! There are an enormous number of radio waves which flow unimpeded past aerials which are not tuned to receive them. Even then, what we hear is conditioned by many things. Hearing a known language is very different sensation to hearing a foreign one; music to one is a cacophony to another. The same is true of sight. Script in a language we can read appears differently to one we cannot, and very differently to one in an unfamiliar one, be it Cyrillic or Chinese if we are Europeans or those scripts to Eurasians or Chinese. The touch-sensitive nerve cells in our skins respond directly to pressure, and the taste buds on our tongues convert the chemicals associated with taste into electrical pulses as do the olfactory ones in our noses which are similarly chemoreceptors. As with all sensations, the electrical pulses travel along the relevant nerves to the brain, within which another hugely complex reaction takes place and we see, feel, hear, taste or smell.

By and large we all sense similarly, although this is difficult to prove except by common sense. The colour red which I perceive is the same as your perception. The justification for this is the almost identical nervous structure which we all share. There are of course some exceptions, as in the neurological disease, synaesthesia, previously mentioned, where the senses get muddled up and sounds are seen or smells heard.

Whether a rose is perceived as fragrant or noisy is an accident of evolution and there is no sense in asking, "But which of these is true?" It is undeniably true that the great majority of us will perceive it as a fragrant flower. It is a worthless speculation that had the course of evolution been different the great majority of us might perceive it as a cacophonous smell with ill-defined boundaries. If our only sensory apparatus was the detection of gravitational forces the boundaries of all objects would be vast.

A favourite answer of politicians when asked the policy of dealing with some adverse circumstance is *I do not deal in hypotheticals* and the same will apply here. The world is as it is[23] and speculations as to what it might have been, or what it would be if different are just that: speculations and vain.

The sensations themselves, sights and sounds and the rest, are accompanied by feelings of joy or sadness, wonder or tedium, pleasure or pain, relief or frustration. These feelings are spontaneous and are the product of our nervous systems, the result both of our genetic heritage and of our nurture. It is possible to trace and understand the precise neural activity which promotes the feelings of pleasure or pain: the firing of groups of neurons associated with the response to a rhythm, for example. Some rhythms appeal universally to humans, to some animals and possibly even to some plants and these can be a subject of research and codification. We can learn a great deal as to which are pleasurable; why they are is a much more difficult problem. We can understand what gives us pleasure but when asked why this is so can only answer that that is the way things are.

An evolutionary answer is that for the survival of any species feelings of pleasure or pain have proved beneficial. The stroking of a cat under its chin, which causes it to purr and is obviously pleasurable, whereas standing on its tail causes it to yowl; both doubtless confer some benefit.

Some of the most powerful sensations, such as the satisfaction which comes from the release from tension following the dawning of an insight into a problem which has been occupying us for many hours might not be the result of immediate sense perception but are due to the way our physical structure is formed.

To sum up: all knowledge comes to us via our five external senses and those internal ones which can transmit pain or pleasure. Sense impressions reach the brain in the form of electrical pulses along the cranial nerves. The receptor neurons in the brain distribute these along an unimaginably vast array of synapses to memory cells, to cells which in turn transmit pulses down nerves to various bodily activities

---

[23] cf The opening words of Wittgenstein's Tractatus Logico-Philosophicus: Die Welt ist alles, was der Fall ist. The world is everything that is the case.

such as muscle contraction, to cells involved in thought and to cells which awake consciousness of sight and sound, taste, touch and smell. There is, of course, a vast amount of innate 'knowledge', such as how to walk and talk, which is due to the evolutionary process but this essay is not concerned with such.

4   Inside the brain: the perception of physical objects.

A human brain has in the order of one hundred billion cells, neurons, connected to each other by one thousand trillion[24] synapses and glia, complex threadlike structures and specialised cells forming a neural net of unimaginable size and complexity. There are several distinct regions in the brain. To get some idea of the complexity, just one of the regions is the grey matter, a layer of between two and three millimetres thick surrounding the white matter in the brain. It consists of some two hundred million columns of cells, each two to three millimetres long, containing up to one hundred neurons. Extensive research into the structure of these columns and their function has been carried out for the last twenty years by the Ecole Polytechnique de Lausanne, among many others. They recently (March 2023) likened the research to an examination of an area of woodland: trees, branches, leaves, roots and the soil, millimetre by millimetre!

The whole is a living, constant whirlwind of activity, operating at relatively low speeds compared to computers, but in parallel in vast numbers enabling all the hugely diverse bodily functions to be coordinated and controlled. These include thoughts and movements, digestion and excretion, generation, birth, growth, ageing and finally, when control breaks down, dying.

Even after the raw data has entered the brain a great deal of processing goes on, as is evident from the different reactions both in individuals and in others. Whereas a piece of Bach might be delightful to an individual in some moods, in other circumstances it can be tedious; hard rock is music to one person (or a hundred thousand at Glastonbury) and a hideous racket to at least one known lover of

---

[24] A billion is a thousand million; a trillion is a thousand billions.

Bach. It is not all that unusual, particularly in a confrontational situation or where a raw nerve has been touched for there to be a marked discrepancy between words spoken and heard. The raw unperceived perception is coloured by many factors, all involving short term memories and those stretching back to childhood and before. Memories in the broadest sense are laid down as soon as the brain begins to form – seconds after conception. Memories are both conscious and unconscious. They are necessary for the simple functioning of a body so that it can digest food, walk, think. Indeed, after suffering major trauma it is commonplace to hear, "She had forgotten how to walk and had to learn all over again," or "He can't remember how to speak."

Virtually every activity, human. mammalian, reptilian, avian – in all living creatures – depends on memories stored in the creatures' brains or in the case where there is no brain, in their nervous systems.

The physical structure of memories requires some permanent or semi-permanent alteration to neurons and the synaptic transmissions via glia between neurons. In the case of those of which we are not conscious, very little is as yet understood although there is an abundance of papers citing theories based on suggestions of the involvement of electrical potentials in glia. How these unconscious memories are accessed is even less understood, and as to how they enter the realm of self-conscious awareness there is no understanding whatsoever, not a glimmer of understanding.

It is often insufficiently realised that all human activity, including thought, and even abstract thought, must have a material element, be accompanied by alterations in cell structure and hence in theory available to external observers while the thought is occurring and so long as any memory of it remains. Farewell privacy!

And as an aside, and the subject for a later chapter, the belief in an afterlife needs to take into account the destruction after death of all physical memory, which is a necessary component of all human awareness. What is man in heaven if he is without memory? Not much point in someone else going to heaven for my virtues or for my vices being punished in hell!

After the brief introduction to current scientific belief, intended to give some idea of the complexity and shadowy nature of the components of every material object, the very difficult concepts of space and time, and the transmission from our senses to our brains, and the functioning of the brain itself, we can start to give a partial answer to the question of our knowledge of material objects.

Well, for a start, we do not say that we know a billiard ball. To the question, "You know a billiard ball?" we say we know what a billiard ball is, meaning in everyday use that we know that a billiard ball is a small smooth solid spherical white or red ball used for playing a game. The appearance and feel of it is stored in our memory and produced usually instantaneously and without thought. If it is not a very common object some effort might be needed to bring its image to mind, and if this is unsuccessful the answer has to be, "No!" As far as this chapter is concerned the only knowledge of interest has come from actually seeing and handling a billiard ball. To be reasonably sure that a particular object offered for identification is a billiard ball it would be necessary to handle it so as to feel its smoothness and weight and have to have handled one previously. Even then it would be difficult to be sure, as the ordinary citizen will not recognise the difference between a billiard ball, a snooker ball or a pool ball from mere handling of any of them: they are very similar. Knowledge from seeing one on television or in a picture in a book, or even less, from a description in a book, is a future topic, and the difference between billiard balls, snooker balls and pool balls will then be discussed. But what is a billiard ball? It is a very smooth hard dense sphere with a diameter of between sixty-one and sixty-one and a half millimetres and a weight of one hundred and forty grams.

When such an object is correctly identified knowledge is certainly shown, but very limited knowledge. If asked, "Yes, but what is a billiard ball?" a whole host of questions could be posed: "Yes, I know it is a billiard ball but that doesn't really tell me anything apart from a name. What is it used for? What is it made of?" and then to how ever many answers to this last, further questions and further: "What is a plastic? What is a molecule? What is an atom? What is a proton? What is a wave?' And so on and so on for ever and ever.

We can only know material objects in the first place by experiencing them ourselves and being told what they are, storing the information in our memories and accessing this as needed and then only giving a series of names or, at a second remove by reading about them or seeing pictures, in the way we learn to distinguish various birds and flowers. At a rather deeper level, we observe patterns of behaviour and call them laws, or more probably read about others' observations and try to understand the implications of the laws.

We know material objects in much the same way as we know a language. We recognise them as we recognise the import of the sounds of a foreign tongue: their images are embedded in our memories.

This is what is meant when we say we know something.

A large proportion of humanity, some billions of people, are taught by their priests and philosophers that our perception of matter is illusory – whether the great majority of them believe this is another matter.

The Eastern beliefs that all material objects are illusory is a central principle of several philosophical and spiritual traditions that originated in the Indian subcontinent. Chief among them are Hinduism, Buddhism and Jainism; they are based on the concept of Maya, which refers to the illusionary nature of the material world which we perceive through our senses.

According to Maya in Hinduism, the world is an illusion, a play of the supreme consciousness of God. It is a projection of things and forms that are temporarily phenomenal and sustain the illusion of oneness and permanence. The illusion of the phenomenal world is created and sustained by stand-alone objects thrown together either by an act of randomness or through the deliberate choice of conscious will. It is the illusion which keeps us from realising the true nature of reality.

The concept is also central to Buddhism, which teaches that everything in the material world is impermanent and subject to change. According to both Hinduism and Buddhism, attachment to material things is the cause of all suffering; the only way to escape is by the complete detachment from the material world which leads to enlightenment and the realisation of the true nature of reality.

In Jainism, the concept of Maya is closely related to the idea of Karma, the principle of cause and effect. It is the belief that actions in this life determine our fate in future lives. Attachment to the material world is one of the main causes of bad karma, which can lead to rebirth in lower forms of life.

Belief in Maya is not a denial of the existence of the material world but rather an understanding that our perception of the material world is not the ultimate reality. The material world is seen as a manifestation of a deeper, spiritual reality which is beyond our senses and intellect

Although the concept of Maya is anathema to the western scientific mind, the four sections which make up Part 1 of this work might make for some sympathy towards it. Before this short diversion into Eastern thought it was said:

We know material objects in much the same way as we know a language. We recognise them as we recognise the import of the sounds of a foreign tongue: their images are embedded in our memories.

This is what is meant when we say we know something.

What about our knowledge of persons. Is this in any way different?

PART 2

THE PSYCHE: Knowing living creatures

Is there an essential difference between the way we know inanimate objects and animals? And then, between knowing an animal – say a pet dog – and knowing another person. Is it merely a matter of extent? In what way do we know ourselves? Is this different from the way we know other persons?

As far as persons are concerned, it is true that there are degrees, varying from the mere recognition that the objects approaching from a distance are people and not a herd of cattle, through knowing that the person greeting you is a Mr Smith, to the recognition that it is a close friend. The addition of a name adds very little apart possibly from an avoidance of embarrassment: we might know virtually nothing else about him. The knowledge is of the same order as knowing that an ash tree is an ash. It is possible, indeed very common, to have work colleagues for many years, to be extremely aware of their moods and phobias, their health and selected political views and yet hardly know them at all. This is similar to knowing that the ash in the park across the road sometimes comes into leaf before the oak (triggering the superstition that we are in for a soak), has seeds shaped like little propellers and sheds its leaves in winter.

Husbands, wives and lovers, close friends – what is the situation here? Again, there are degrees. To a much greater extent than with work colleagues we know their moods and foibles, their views on a great number of subjects; we know that we love them and that they love us. At least, we have no doubt that this is the case. Can we be absolutely sure? We have all had friends or heard stories about the shock of spouses discovering themselves to have been grossly deceived? Could this possibly be the case in our own case? Of course not.

1 Animals

Does the way in which we know animals differ in any way from the way in which we know inanimate objects?

Homo sapiens is a product – the culmination, some think – of the course of evolution which started with the formation of the universe, thought to be around fourteen billion years ago with the Big Bang

Before this there was no before, no space, no time, nothing, nothing whatsoever. The tortuous route of evolution from elementary particles through the formation of molecules, DNA and living creatures has been sketched previously and it is clear that as one moves along the evolutionary tree each product has a great deal in common with its ancestor, and close to an identical structure to its immediate parent. Thus, in the phylum Chordata, a blue tit is closely related to a great tit, and a wolf to a dog. Humans are related to all other mammals remotely through common ancestors, more recently to the primates, and almost identically to their parents.

All the emotions which humans feel – love, hatred, pride, happiness, misery and despair – are exhibited by animals. Some are obvious and some can be deduced. A kitten playing with a ball of wool, the disdain of adult cats, affection and jealousy of dogs, rage of a bull, the evident joy of gambolling lambs or baby otters sliding down a mud slope, who has not witnessed at least some of these? The aggression of a robin when another disturbs its sweet carol, the repose of a swan idling on still water turning to fury if a person comes to close, the vanity of the winning racehorse. All these are the result of evolution, and since all animals have come from common ancestors, and very likely from a single common ancestor, we can deduce that the various emotions are common to all, in the degree to which their nervous systems allow. We are not privy to the dreams of an earthworm but cannot doubt that they exist in some vestigial form.

When Fifi jumps for joy when her owner returns from a short absence, or assumes a hang-dog expression when caught stealing a piece of cake is she aware of what she is feeling, *Naughty Fifi, naughty naughty dog*? There is no simple answer to this. The degree of conscious awareness in living animals might vary all the way from human self-consciousness to that of a digital computer, which has none whatsoever. It would be quite possible to construct a computer which would purr if all its systems were in good order, or whine if it was running short of charge, but computers have not a vestige of what is generally meant by 'consciousness'. For all the extraordinary advances since digital computers emerged in the middle of the twentieth century, almost a hundred years ago, they have not advanced in self-conscious awareness at all: they have none, none whatsoever. Word is a very sophisticated word-processing application, and although in some sense the computer which is being used to type this document can be said to be aware of what it is doing, it is certainly not self-aware. Animals are hugely more complex structures than any current computer, and their awareness much more sensitive, but it cannot be ruled out, it is not unimaginable, that they have no more self-consciousness than that of a computer.

If they are not self-consciously aware, then the extent to which anthropomorphism influences theories of animal cruelty need examining.

Take battery hens. The hens are confined close to each other in conditions in which they never see natural light or have any freedom to roam and scrabble for their food; they lose their feathers and exist in an almost naked state: inhuman conditions, we say, really cruel. On the other hand, in some ways they absolutely thrive. They are safe from foxes, very well fed, healthy apart from unnecessary feathers, live comparatively short but stress-free lives and lay eggs galore.

Horses, dogs and donkeys have been used by mankind for tens of thousands of years, cattle and goats have been a source of food and milk; cattle, goats, sheep and bears have also been bred for their skins, wool or fur. The great majority of humans lived for tens of thousands of years in much closer proximity to animals before the universal drift to cities since the time of the Industrial Revolution and it might be thought that they knew more about animals than today's city dwellers whose only close contact is a pet. This is not to say that the animal rights movement is wrong: it is to say that the assumption that animals should be treated as quasi-humans is not an obvious moral imperative, unlike cruelty, which is. Whereas Shakespeare's wanton boys mentioned earlier are clearly guilty, hunting, shooting and fishing are not obviously so.

With respect to self-conscious awareness there are two possibilities: either matter has some element which allows self-consciousness to emerge through evolution alone or it does not. This is the subject of panpsychism, the belief that everything material, however small, has an element of individual consciousness. Elements of it are found in the Pre-Socratics. Pythagoras, early 6th Century BC and his followers, gives an almost religious importance to numbers and their relation to physical objects. This philosophy profoundly influenced Anaxagoras, who came a little over half a century later. He believed all matter to be almost infinitely divisible with the fundamental units containing all the elements including thought; he gave them the name of Nous, (mind, or intelligence).

If all matter is potentially conscious (in the sense of self-conscious awareness) this element is entirely unaccounted for in current theories of the nature of matter: it is a matter only for philosophers! It is easy to imagine how with our current knowledge of the nature of matter we will one day come to understand the transformation of inanimate matter to life and come to a greater and greater understanding of the course of evolution. This has been accompanied in some species by a step-by-step increase in nerve and brain function with a corresponding increase in sensitivity: consciousness. It is not possible with our current understanding of the nature of matter to imagine how any living creature could have self-consciousness. There is simply nothing in the constitution of an atom or any other elementary particle, in so far as they are currently understood, to account for it. Of course, it is true that our understanding is far from complete: quantum physics and relativity still present considerable difficulties. The science of life – what makes a living organism tick – has many unknowns. Potential consciousness should not be ruled out a priori.

If, to answer the first part of the dichotomy, it is the case that consciousness and even self-consciousness are fundamental properties of matter we cannot escape from the thought that not only all – *all* – living creatures have some degree of consciousness, but every particle of matter too. Any arbitrary classification of thresholds of possibility of feeling pain – this animal can feel it, this cannot – become untenable. All animals, and to some extent, all plants, can experience pain and so, to some lesser extent can stones (but how is pain to be quantified?). The notion of cruelty to some animals needs to be extended to all. Some Hindu and Buddhist doctrines, believing that all life is divine, explicitly endorse this view but it is hard to find much consistency in its formulation, and even harder in its practice. Whereas cows are sacred, dancing bears are not at all. To an ill-informed outsider, pragmatism appears to play a larger part when it comes to the effect of the doctrine on human life than cut-and-dried Western logic would sanction. In most of the West, animal vivisection for medical purposes is legal; battery hens became unlawful in 2013. In Europe there is a bewildering set of laws. Bull fighting in Spain has a large following but an increasing opposition; fox hunting in the UK is illegal but grouse and other game-bird shooting is allowed. It is illegal to trap and drown a grey squirrel (a really quick death) but quite all right to trap one and condemn it to several hours' of very agitated existence in the trap while taking it to a vet or NSPCA for slaughter.

If all animals can feel pain, is cruelty to be defined by extent? If so, who is to be the arbiter? At present, in the enlightened West, the threshold is set at the possession of an advanced central nervous system, and the arbiter is public opinion. But public opinion is vague and delegates responsibility to elected representatives who nearly always enjoy the support of less than half the population. It forms the laws which the public must follow – or repudiate at the next election. This is unsatisfactory from a logical point of view, and the discussion of its implications will form part of Part 4: Right and Wrong.

Where does all this leave us in trying to understand what we mean when we say we know an animal, beloved Fifi, for instance? She clearly recognises her owner when she sees her – indeed, very often long before she comes into view: the noise of the car coming up the drive is sufficient to send her into paroxysms of delight. Mutual recognition between animals and humans is very common and to this extent they may be said to know each other. Inanimate objects are known by having their images in our memories together with a certain amount of information about them. Animate objects are known in the same way but sometimes with an additional element: Mr Jones knows that Fifi is a poodle; Mrs Jones knows that that it is a poodle, her very own poodle which knows and loves her in return.

The second half of the dichotomy, that there is nothing in the nature of matter alone which allows for evolution to result in the emergence of creatures with self-conscious awareness will now be considered. By alone is meant without some additional non-material addition. This leads to an examination of what constitutes a person and the nature of mutual knowledge.

2 Persons
2.1 Self-Knowledge.

γνῶθι σε αυτόν[25]

O, wad the Power the giftie gie us
To see oursels as others see us![26]

A good place to start this investigation is by looking at the person who perhaps is known best of all: oneself.

What do we mean when we say that we know ourselves? To say that we recognise ourselves or have images of ourselves in our memories is more or less meaningless. We are aware that we are unique individuals. *What am I?* Kim agonises.[27] From our earliest moments of self-consciousness, at the age of two or three, we have daily but not unbroken experience of ourselves. Memories of early childhood are sparse, and the age of earliest memories varies considerably from person to person. Many early memories are not memories at all of the event but of what we have been told what happened; one reads of true memories of early happenings of which there was no or very limited contemporary consciousness being revealed in dreams many years later.

Before listing matters about which people can be confident that self-knowledge is factual, such as physical features, there are a number where they are very much in the dark.
In particular, three stand out: motives, and secondly, coupled with this, judgement of rectitude, and lastly freedom, free will. In all three of these there are two means of obtaining knowledge. The first is introspection, possibly assisted by sessions on the psychiatrist's couch and the second by being told by those close to one.

Man's proclivity for self-deception is almost unbounded; this varies from believing all one's actions to be virtuous and altruistic, to the

---

[25] Know yourself: Inscribed on the pediment of the temple of Apollo at Delphi.
[26] Burns
[27] Kipling, Kim Chapter 15

conviction that one's every act is worthless, base and self-seeking. It is extremely rare for either of these states to be even remotely the case, and very rare for self-judgement to be accurate.

2.1.1. Motives and Self-Judgement

It is an inescapable predicament of the human condition to be prone to self-delusion but occasionally we are forced to recognise some error in our self-esteem and occasionally the opposite. There is a character in Steinbeck's *Sweet Thursday* who thinks worse of herself than she should, who is advised by her Madame:

> *First you got to remember that you are Suzy and that you*
> *ain't nobody else but Suzy. Then you got to remember*
> *that Suzy is a good thing – a real valuable thing – and*
> *there ain't nothing like it in the world. It don't do no*
> *harm just to say that to yourself. Then when you got that,*
> *remember that there's one hell of a lot Suzy don't know.*

On the other hand, one of Eliot's gifts reserved for age is

> *Motives late revealed and the awareness*
> *Of things ill done to others' harm*
> *Which once we took for exercise of virtue*[28]

"I surprised myself," for good or evil is often heard. Is self-knowledge essentially of the same nature as knowing others? It is of course obviously true that we are much more aware of ourselves than we are of those around us – would at times that we were less – but the nature of our observing is essentially the same. We have privileged access only in so far that we are, as it were, to hand.

Our self-knowledge is privileged in so far as we are always present; it is hampered by the constant temptation to make excuses for our behaviour; it is limited by the location of some of our organs (there is no way that one eye can directly survey the other), or the inflexibility of our bodies so that to see the top of our head is impossible without a mirror. On the other hand, I smell myself in much the same way as I smell you and regard my finger nails as I do yours but more (or less) critically.

When it comes to having a headache, only the sufferer can actually experience it. I feel my headache in an entirely different way to feeling yours – in fact I do not feel yours at all. A person can only

---

[28] Little Gidding

experience a headache when certain cells within his brain fire and this is purely a personal experience. It is also a strange fact that I can have a headache and be distracted by something interesting and then temporally forget that I have one until the distraction ends and the ache returns. The ache has certainly gone but has there been any change in the neurons that cause the ache?

It is true that one may be aware of facts only observable to others, such as keeping hands behind one in the knowledge of having dirty nails!

Too much introspection is probably not a good thing as it leads to rather too much absorption in self, but a little – a nightly examination of conscience, perhaps – can be salutary[29]. It can be a useful exercise to imagine yourself in front of a stern judge, think how you have wasted your life, ignored the plight of others, been petty, mean and selfish and, and .   .   . there is no end to it: you stand utterly condemned. Even the few good acts which can be summoned are shown to be self-serving. Then appearing before a lenient and compassionate judge and learning how much can be excused, how you didn't really have a choice, were unfairly tempted, didn't mean to do it, and so on and so on. Not only can we not judge others (we never know all the circumstances), no more can we judge ourselves.

2.1.2   Freedom of Action

In deciding whether any action has been freely performed introspection is not very helpful. Everyone has experienced coming to a decision, after much deliberation, "Right, I will do it," but the thought process is far from clear and the acting on the decision very difficult to pinpoint.

There are two major problems with the notion of free will. The first is in coming to some understanding of the agent of choice, and the second lies in the necessity of interfering with the laws of physics.

---

[29] As an aside, Matthew Parris in his column in the Times recently wrote that although he was an atheist he was brought up to say his night prayers, and has continued this as a useful daily check all his life. He dictates all his work into a Dictaphone, taking great care to spell out the punctuation. He caught himself praying "Please grant that my partner's show is successful. Stop."

2.1.2.1 The agent of choice

Any human (or indeed animal) action is the result of the firing of huge numbers of neurons sending and receiving pulses along nerves between brain and muscles. This process is an essential part of living. It is the result of aeons of evolution and in theory is susceptible to straightforward scientific analysis. How then can it be possible that the firing of some individual neurons is directed to act in a way that differs from their normal behaviour? Who does the directing? And how?

It is clearly utterly absurd to imagine that there is some 'Me' within my head controlling individual neurons.

It is only in human actions in which there is held to be an element of free will that this problem arises.

To counter this argument, we need to consider what the agent is in any human action, and we come up against the problem of self-consciousness. When I am aware of something, who or what is the 'I' that is aware. When I stub my toe it is not the toe which feels the pain. *I* feel the pain in my foot. The toe is only part of me; lop it off, and I am still me. Make me a quadruple amputee, I am still undeniably me. There is a story in Raoul Dahl's Kiss Kiss where a bullied wife manages to persuade a hospital to keep her deceased husband's brain and one eye alive, by supplying it with blood as necessary. She puffs cigarette smoke into the eye, which blinks in fury. This is a macabre imagination's science fiction idea of the irreducible essence of a human being. It takes us no nearer to any understanding of how we might have any control of our actions. We might agree with Freud's analysis in which, in simplistic terms, the ego is the conscious mediator between the struggle between desires and fears, and conscience, the id and the super-ego, but this does not address the mechanism of the ego. There might be a glimmer of the possibility of an understanding, in Aquinas' refinement of the Aristotelian theory of matter and form where the form of a human is the soul, which informs every action. This is further discussed below in Part 5.1, The Notion of Ensoulment. The theory entails the belief that every part of every human, down to individual atoms, has as its form a non-material element, the soul. To say that this is a difficult concept is to put it mildly. As with quantum theory, there may be those who understand it or, as has been quoted of Richard Feynman, there may not!

2.1.2.2 Interference with physical laws

A more serious problem with the notion of free will is the necessary interference with the chain of cause and effect, in accordance with the laws of physics, or, at the elementary particle level, with the statistics of quantum physics. A free choice necessarily involves the alteration in the state of brain cells, leading to movement of muscles, which would not have occurred otherwise. If this is not the case, the thought that the action was free is an illusion. Even if the involvement of choice is no more than a mental acceptance of some behaviour ('All right, I will do it.", when you would have done it anyway) is ultimately no more than an illusion of freedom. The main difficulty in defending free will is the agent. Who makes the decision? There is no 'Me' residing somewhere within me giving instructions – the idea is ridiculous, has no evidential support and is contrary to what we know of ourselves by introspection.

All manner of arguments are advanced in denial of anything being needed apart from the atoms out of which the brain is made. They all involve the thought of many desires coursing through the brain and one of them dominating, and all maintain that ultimately free will is illusory. The problem with this is that virtually no one believes this. Without it there can be no responsibility for any action, any more than a cow has for choosing to graze this bit of grass rather than that. It is difficult to deny that this might be the case, but if so it is profoundly depressing. It reduces the whole gamut of human civilization, its efforts at law making and ethical teaching, art and literature, to a mockery. These thoughts are not the main purpose of this present chapter: they will form the main matter of Part 4, Right and Wrong. For the moment let it suffice that our self-knowledge in respect of our fundamental freedom is indeed very cloudy!

Descartes in searching for a rock of belief of which there could be no doubt, settled on self-knowledge in his famous *Cogito ergo sum – I think therefore I exist.* Although it is theoretically possible to be mistaken in any given sense perception – what I think I am looking at is in fact an illusion – this is an extreme position or at least an unnecessary one. We can be quite sure that except in very rare circumstances our perceptions are not illusory. When I see a chair, I see a chair and not a hologram.; when I see a pink elephant floating past . . . . The exceptions might be due to deliberate fraud or more likely to an illness but they are rare and there is not much sense in founding one's philosophy on them. Much better is to assume that they are rare and that sometimes we might be mistaken. Apart from being aware that we exist, which is not of any great value, there is no sense perception of which it would be absolutely impossible that we might be mistaken.

As far as knowing ourselves is concerned our knowledge is only marginally more privileged than our knowledge of others, except in so far as we sometimes know what we were thinking about and never *know* what you were thinking about. We know ourselves by straightforward observation. We observe our appearance, our behaviour, we hear what we say and might remember it. We do not observe our thoughts: we think[30].

It is a misuse of language to say that what we know most directly are our own perceptions since to know our perceptions would need a knower apart from them, an I within me. This leads to an infinite regression: if we could perceive our perceptions we could perceive our perceiving them. . . .

Perceiving is an act of consciousness involving an internal or external stimulus. We can perceive light or sound, an ache or a pleasant taste; we do not perceive thoughts or memories: we think or remember!

2.2   Other Persons

Adam knew Eve and she conceived and bore a child.

One could write reams about the biblical use of the verb *to know*. The words used in various versions of the Old Testament, Hebrew, Greek and Latin and the import of the choice of the particular one. (Latin has at least four: scire, noscere, cognoscere, agnoscere; English has dozens: know, perceive, admit, comprehend, realise, discern, be conscious of, be conversant with, be aware of, understand, be acquainted with. . . .).

As far as this essay is concerned, the biblical use is mentioned only in so far as it can, in rare circumstances, be the most intimate knowledge one person has of another.

In the poem below, Donne writes of an out-of-the-body experience which he claims was not sexual, and it is the possibility of the truth of this statement which is the central issue with this section.

---

[30] cf Wittgenstein Philosophical Investigations ll xi Blackwell 1963 p222, is in partial agreement: "It is correct to say 'I know what you are thinking' and wrong to say I know what I am thinking.' (A whole cloud of philosophy condensed into a drop of grammar.")

Where, like a pillow on a bed,
A pregnant bank swelled up to rest,
The violet's reclining head
Sat we two, one another's best.

Our hands were firmly cemented
With a fast balm, which thence did   spring,
Our eye-beams twisted, and did thread
Our eyes upon one double string:

So to 'intergraft our hands, as yet
Was all our means to make us one
And pictures in our eyes to get
Was all our propagation.

As 'twixt two armies, Fate
Suspends uncertain victory,
Our souls, (which to advance their state,
Were gone out), hung twixt her and me.

And whilst our souls negotiate there,
We like sepulchral statues lay;
All day, the same our postures were,
And we said nothing all the day.

This ecstasy doth unperplex
(We said) and tell us what we love.
We see by this it was not sex,
We see, we saw not what did move

When love with one another so
Interanimates two souls.
That abler soul, which thence does flow,
Defects of loneliness controls.

We then, who are this new soul, know,
Of what we are composed, and made,
For th'atomies of which we grow,

Are souls, whom no change can invade[31]

When we say that we know someone, what do we mean? Is there anything to add to, different from, knowing about them? Is there anything in that knowledge that sets them aside from other living creatures?

Is a human being unique or just one mammal among others? The recent evacuation of horses in a war zone[32], thought to be in preference to men, women and children, caused widespread moral outrage. It is generally agreed that man has a 'higher' status than animals, not in the sense that a neurosurgeon has a higher status than a milkmaid but something more fundamental. It should not need repeating that a general agreement that something is right carries no guarantee of its rectitude! It is indisputable that until modern times the placing of man above dumb brutes was unquestioned. Perhaps it should be – no, it certainly should be.

We know we have self-conscious awareness; we believe we have individual freedom and that we can use this for good or ill. We do not know if the first of these is true of the rest of the animal kingdom but we do not believe that any of its members can act in morally good or evil ways.

The more we consider the matter, the more it becomes clear that when we say we know another person what we mean is that we recognise them, that they are on our memories in many moods and situations. Self-knowledge and knowledge of inanimate and animate objects is all of the same nature: the difference is in the object known. The only difference in knowing a stone and knowing a tree and knowing a dog and knowing a close friend and knowing oneself is in the objects of the knowledge..

What then of the experience of John Donne in his Ecstasy? He says in the poem that for all that their hands were firmly cemented, intergrafted, all the day, it was not a sensual experience but a true meeting of souls? This form of expression is more common in mystical writers than erotic ones. There is no hint of it in his sermons or in the later Holy Sonnets and Divine poems. Rather than posit a meeting of souls, whatever this might mean, it seems much more likely that it was an instance of an extraordinarily heightened

---

[31] From The Ecstasy, John Donne
[32] Kabul, August 2021

awareness, not all that unusual in the early stages of passion. These are completely sensual and our understanding of them depends on our understanding of the nervous system, in the same way as any other human emotion. "Completely sensual" does not exclude that possibility of humans having a non-material component but it does cast doubt that only a non-material element, the soul, is involved.

Even though we only know about other persons in the same way as we know about any material object, it is indisputable that with the rare exception of a very few people in respect of a very few higher primates, what we know about other persons is that they are members of a species quite different from dumb beasts. We recognise them as being able to think rationally (up to a point), to be able to act rightly or wrongly and to bear some responsibility for their actions. We are members of the same species and have a great deal in common with all other members. Our knowledge of them is partly through personal observation and partly – and probably more importantly – from observation of ourselves. The intention of the designers of the Delphic temple was probably a recommendation to be aware of one's limitations, and not the sense here suggested: it is coupled with another inscription advising moderation. Nevertheless, the more we recognize the wellsprings of our own behaviour the better will we understand that of others. Both are fraught with difficulty and in neither can anyone be confident that the knowledge is sure. We cannot, as has already been said, judge either ourselves or others with any confidence.

Is there some way in which the human species, homo sapiens, is fundamentally different to the rest of the animal kingdom? Has man something in addition to material atoms rather than material atoms having some quality as yet unknown to science? If it exists, this something is known as soul: a non-material spiritual element. Each person is held to be in possession their own unique one, through which they have the power to reason and which confers on them immortality. The difficulties of this concept will be discussed in Part 5.1, The Notion of Ensoulment, but as far as the intercourse of humans is concerned there is no reason to believe that this differs in any way from every other human action, in which the whole person is involved. There are well authenticated happenings, such as telepathy and seeing future events, but these too involve the complete person and if there is any substance in the emerging theories of block time they do not require the belief of independent spiritual activity.

It is a probable (and regrettable) truth that I love my poodle, Fifi, my mistress, Fifi, and my wife, Fifi, (my bit-bull Fido, my lover, Fido, and my husband, Fido) in exactly the same way with the exception that, pace Freud, there is probably little of the sexual in my relationship with the first Fifi or Fido.

It seems unimaginable that when I say 'I know my dog,' my mistress and my wife, the word *know* is being used in exactly the same sense and has exactly the same meaning.

Is the knowledge that we have of ourselves any different or only greater because we have more opportunity of observing ourselves?

This section started with a poem and will close with one which would perhaps be better placed in Part 5. Yeats must at some time have wrestled with this intractable problem: knowing directly, rather than knowing about or knowing that.

> *Never shall a young man,*
> *Thrown into despair*
> *By those great honey-coloured*
> *Ramparts at your ear,*
> *Love you for yourself alone*
> *And not your yellow hair.'*

> *But I can get a hair-dye*
> *And set such colour there,*
> *Brown, or black, or carrot,*
> *That young men in despair*
> *May love me for myself alone*
> *And not my yellow hair.'*

> *I heard an old religious man*
> *But yesternight declare*
> *That he had found a text to prove*
> *That only God, my dear,*
> *Could love you for yourself alone*
> *And not your yellow hair.*[33]

---

[33] Yeats For Anne Gregory

In short:

The course of evolution from unstructured elementary wave-particles at the beginning of the universe to today's immense diversity is seamless. This encompasses all living creatures, including self-conscious ones. It has no discontinuities. It is the result of the accumulation of quantum (largely electron-photon) interchanges. Self-conscious awareness aside, this is an entirely intelligible process and there is overwhelming evidence, from many different disciplines, that this is in fact the case. If it is wholly a material process, then self-conscious awareness must be a fundamental property of matter, and although there is little evidence that this is the case it remains a possibility. If it is not a fundamental property of matter, then man must have a non-material element, conferring on him the ability to reason, some true freedom and responsibility for his actions, entirely absent from all other species.

Freedom and responsibility have been touched on; the ability to reason will now be examined.

3 Clouds of abstraction

If any concrete objects are taken – it might be a field of cows, (some years ago, really concrete in Milton Keynes) an avenue of trees or a group of kangaroos. Count their numbers: fifteen concrete cows, sixty-four trees, nine kangaroos. Numbers are not particular to cows, copses or kangaroos – they are, as it were, stand-alone notions: abstractions. Consider, for instance, our objects' colours: white and black, green, brown. Now consider the colours but not what is coloured: colours in the abstract. Or weight or feel or smell. Consider a just person and forgetting the particular person: think about probity in the abstract. We have the capacity to take away the particular in any observed quality or quantity. This is not confined to humans. Animals from larks to labradors can be trained to recognise numbers, colours and shapes but probably not weights; they can separate appropriate blue objects from green ones. And any self-respecting computer can do all this, of course.

We do not know any material object directly: we only have knowledge about it. The situation with abstract knowledge is to a certain extent quite different: We really do know that two and two make four, that green is restful to the eye, and we think we really do know that totalitarianism is an evil form of government. In the domain of ideas, we have direct knowledge. Now, a dog can be trained to know that two and two make four (*Fido, what are two and two? Woof, woof, woof, woof*) and a man can be similarly trained to know that an iambus is the same in Latin and English, without the foggiest notion of what an iambus is. It is more than possible to get a reasonable grade in A Level maths, having learnt among other things that the integral of x squared is one third of x cubed with precious little understanding of why this should be the case! Rote learning does not add to one's knowledge in any substantial way – much as learning a foreign language does not, although this does provide a useful tool and does give access to a foreign literature in a way that no translation can. In a structured language, it can lead to an understanding of the rules of grammar in addition to just knowing them.

There are degrees of abstraction. For example, a cow. Long before they are able to read, children can recognise the shapes of many animals and their sounds. Then there are further abstractions: the notions just of shape, of sound alone. In the former there is an element of knowing about, prone to error. (We know of a mother who was playing animal sounds to her two-year old, showing her pictures:

Firstly, a cow? *Mooo!* said the child Then a cat? *Miaow!* Followed by a dog? *Woof woof!* And then a gorilla? *Mummy!)*

The idea of no particular shape at all, just shape is more difficult. There can be no element of doubt here. We know what is meant by shape, even if the notion is hazy, in much the same way as we know what energy is, or matter without any shape whatsoever. Does a computer know these absolute abstractions?

It is probable that all living creatures have powers of abstraction but very unlikely that they know that they have. This knowledge can only be gained by creatures with the ability to reason, which in turn requires the possession of self-conscious awareness. Although it is possible to reason unconsciously, as during the night the solution to some problem is found when asleep (particularly crossword clues or some theory we find difficult to understand), ready to hand when awaking in the morning, this only becomes knowledge with the conscious awareness of it.

Computers can count – indeed the one on which this document is being typed counts all the time; records are shown at the bottom of the screen giving the number of pages and number of words. Computers can solve crossword puzzles. Take an example from The Times several years ago:

> Clue: *Oiled silk? Stuff! 5,2,1,5*
> Computer process: *Oiled, 8th definition in its online dictionary: See Well-oiled*
> *Well-oiled: Drunk*
> *Silk (6b) Informal: Queens Counsel*
> *Counsel (4) A barrister, member of judiciary*
> *Judiciary (1) Pertaining to judges*
> *Stuff: (8) Nonsense: therefore not well-oiled*
> *Not oiled? Sober*
> *Common sayings with judge and sober: Sober as a judge 5,2,1,5. Bingo!*

What is the difference between the computer following a programme and the mind casting around for associations? Precious little in the substance of the matter! Both act in accordance with laid-down procedures. In the case of the computer the procedure is a structured programme written by a person; in the case of the person the programme is hard-wired into her brain as a result of evolution. It is less obviously structured and appears to be the result of a more or less random (doubtless less rather than more) search for associations carried out by the repetitive firing of neurons through the billions of synapses that constitute the neural web.

Take another example, the well-known puzzle devised by Edward de Bono:

Take a glass of vinegar and a glass of milk. Put a teaspoonful of vinegar into the milk and stir vigorously. Take a teaspoonful of the mixture and put it back into the vinegar. Which will be the most polluted?

A computer would calculate the percentage of vinegar in the milk in terms of the relative volumes of glass and spoon, even if these are not known. It would then compute the percentage of milk in the vinegar and produce an answer. If we try to the same process we tend to get into a fearful muddle, but there is a much simpler approach to the problem. Since the final volumes of the polluted substances are the same as the initial ones, the milk must have gained exactly the same quantity of vinegar as it has lost milk, and this quantity has been transferred to the vinegar. They will be equally polluted, and the reasoning did not need to be calculated. A simpler example would be to take a bowl of one hundred red beads and another with one hundred blue ones. Take twenty beads from the red bowl and add them to the blue and stir them around. Take twenty of the mixed beads at random and transfer them to the red bowl. However many blue beads this now contains must be the same as the number of red beads in the blue bowl. A computer could very easily be programmed to work this out using arithmetic and a simple algorithm, but to do the same do for the milk and vinegar by calculating the number of molecules of each would be nigh impossible.

For a third example, using Whitehead and Russell's notation

$$P \supset Q = P \vee \sim Q. \quad Df[34]$$

We are interested solely with an understanding of the definition, of what is meant by the various symbols.

> Df means that although it is blindingly obvious it will be formally stated as a definition
>
> $\supset$ means 'implies': if there is some situation p then q will follow. For example, 'If you have a high fever, you have probably got a bacterial infection.
>
> = has its ordinary meaning

---

[34] Principia Mathematica Chapter 1 Cambridge 1910

v is an alternative; ~ is a negation. P v ~Q. means, 'either it is some situation P, or it is not Q .

So the definition states 'if P then Q' is the same as 'either it is P or it is not Q.'

When expressed in symbols this might not appear to be blindingly obvious, but in the type of comment frequently heard, as for instance when your car has broken down, " 'If you were to ask me, I would say it is a head gasket,' is the same as, 'Either I am right or it is not a head gasket.' ".

This equivalence of a conditional statement and a straightforward non-conditional either-or has great utility in computer programming.

The time is almost with us when a computer will write the programme with the only human involvement being the writing of the programme enabling computers to write programmes, with man acting in a quasi-evolutionary role![35]

After much puzzlement the solution to a problem is accompanied by a feeling of elation in the human – another product of evolution. The computer could easily be programmed to exhibit similar joy.

Wherein then lies the difference? The obvious answer is that the computer has no self-conscious awareness whatsoever whereas the human does. Even if computers become sufficiently sophisticated that there is no way that their 'thought' processes can be distinguished from human ones, the situation would not change. In the evolution of computers from the monsters with arrays of glass bulb-like components, taking up whole rooms, to ones with micro-chips smaller than the palm of a hand, containing literally billions of electronic components, there has been no advance whatsoever in self-consciousness, none whatsoever: they are just sophisticated machines. It is of course not unthinkable that an entirely different form of computer will be developed but this is not so far the case. There are various forms of analogue computers, parallel computers and quantum computers using different forms of determinate or statistical programming but still subject to human input and unlikely to be otherwise before a living cell can be synthesised from raw matter. Self-replicating cells are beginning to be created in laboratories but only using living cells or parts of living cells.

---

[35] With the advent of ChatGPT in late 2022 the time seems already to have come.

Is it possible that animals are just sophisticated machines too? Homo sapiens is certainly not. Certainly not?

Is it possible to imagine any fully abstract thought being programmed into a computer? Computers can observe everything in the universe and store the observations in their memories more completely than humans; they can learn languages and argue logically – they can point out flaws of logic in a human argument; they can perform miracles of advanced mathematics, but all of these is a form of 'knowing about.' It is difficult, it is impossible, to think of any human thought-process of partial abstraction which could not be programmed into a digital computer, so there is nothing spiritual needed in their structure. It is very difficult, and possibly impossible, to conceive of any computer being programmed to recognise complete abstractions. It can recognise this shape or that, a triangle or a cow, but it is impossible to present it with no shape at all. Given a definition of *shape* it could say, *That's a shape.* We, human beings, do not need a dictionary definition to form the abstract idea. We merely need to be told that this has this shape and that has that to come to an understanding of the abstract form, shape.

Two observations: the programming must have been carried out by humans; the thought that human beings have evolved from raw matter into creatures capable of programming computers is perhaps more difficult to dismiss.

For each of our senses, through which all knowledge comes to us, there are total abstractions. These are sight (and subdivisions such as shape and colour), sound (including languages and music in the abstract), taste (bitterness and sweetness) and smell and touch. Does it follow that there is something non-material (that is to say, spiritual) in our power of total abstraction?

The possibility that creatures with the ability to have abstract thoughts can be the result solely of natural evolution remains, and the notion of men having souls has not been convincingly demonstrated.

Let us examine a little more closely the idea that raw matter could evolve into creatures capable of abstract thought, capable of constructing computers. The creatures would clearly need to be aware of what they were doing, to be capable of what psychologists call qualia.[36] We are back to the opening paragraph of this chapter and are

---

[36] Ten years ago when *qualia* was typed into Word the spell check asked whether *quails* was meant, although it did not specify whether it was

going round in circles, as tends to be the case when trying to find some understanding of the nature of a soul, a non-material substance, somehow uniquely part of a human being. We have no clear conception of spiritual substances, as we have no means of sensing them: not sight nor sound nor touch nor smell, nor taste. We can infer their existence because no other explanation of qualia is satisfactory., and indeed Plato attempted to do this, not wholly satisfactorily, nearly two and a half thousand years ago. There is another possibility but one which will be scorned by a great number of well-informed, highly educated people: that some of our knowledge has come through our senses by way of revelation by some extra-terrestrial being, a god or angel, or God. It is not the present purpose to discuss the possibility of this; it will be examined in Part 4, Right and Wrong.

The term 'thought' has been used indiscriminately whether by a computer, an animal or man. It was stated in the opening section of this work, that for humans, even abstract thoughts must have a material element. A total abstraction, such as the perception of the concept, *shape,* might in itself have no material element (apart, of course, from the content of the perception) the conscious awareness of it does require neural activity. A possible exception to this would be the experience of being in a trance but this is outside the scope of this essay.

---

thinking of birds or fear. Word has clearly become a little more educated as no such warning is now given.

PART 3

VAST AREAS OF IGNORANCE

1 Knowledge from nurture: home and school.

Opinions need to be underpinned by knowledge to have value. A large proportion of our opinions are formed during childhood and during formal education. Their worth is reliant on the quality of the teaching of parent and school.

Knowing one's table manners are a trivial example of knowing something which can however have serious consequences. The prime purpose of table (and indeed all) manners is to make social intercourse pleasant. We can be taught how to hold our knives and forks so that we 'know' the correct way. The correct way is unfortunately very often merely a way of asserting social class: it should be such that those with whom one is sharing a meal is put at ease. The often quoted though probably apocryphal example of Queen Victoria drinking from her finger bowl to put a guest who had done so at ease is a good example of this. The serious consequence of using manners at odds with those of present company may be alienation.

Ethical opinions are initially formed during childhood and are the subject of Part 4. Political views are formed both from parents and from various forms of media: books, radio, television and newspapers. These, and the knowledge gained from universal subjects of formal education – literature, languages, history, geography, physics, chemistry, biology and maths are the subject of this Part.

Although the quality of the knowledge gained at school varies with the ability of the teachers, it varies even more with the subject. Maths and science at school level is of necessity partial and simplified. History will be influenced by myth and doubtful record, by simplification and by the country in which it is being taught: a Russian account of World War 2 is likely to be markedly different from an American account, and an American one from a British and all of them of their very nature inadequate.

The very purpose of the teaching of any knowledge is not as straightforward as might be thought. It is in part that we become

67

better informed as to the world in which we live, that we might better understand it and our fellows. Equally, and perhaps as important is the training of pupils to be able to use their intelligences more fully. Enhancing memory by practice is beneficial, whether it be of prose or poetry.

2 Media based knowledge.

It must be true. I read it on my tablet.

I beseech ye brethren in the bowels of Christ,
always consider ye might be mistaken.[37]

2.1 Social Media

The most pernicious aspect of media based knowledge is, at present, fake news on tablets and phones used across the globe by children and adults. Children in their billions and many adults too accept the content as gospel. The rapid increase in the powers of AI can make even critical discernment very difficult. Fake text in the reporting of events and situations cannot in itself be seen to be false. The use of ever more accurate mimicry, and the ability to manipulate photos to verify the text, make it almost impossible to distinguish truth from falsehood.

Apart from the corrosive effect of not knowing what to believe, the direct harm from fake news varies from very little to a possibly global catastrophe. In the reporting of the splitting-up of a celebrity couple, apart from the couple themselves, very little harm results; in a tense political situation riots can result and it is not difficult to imagine circumstances in which fake news could start a war, and extremely easy to imagine a nuclear war starting as the result of a false report of one or other nation using an atom bomb.

The safeguarding of free speech, so vital for a free society, makes regulation exceptionally difficult, but some authoritative guarantee of authenticity, which cannot be tampered with, might someday be possible. In the meantime education and extreme caution must be used.

---

[37] Oliver Cromwell, letter to the Synod of the Church of Scotland, 5th August 1650

2.2   Printed material and web-based articles.

A great deal of what we think we know has come from sources apart from our own direct experience. In some of these there is no room for doubt and no reason to doubt. We all know that there are large cities in various countries which we can name – London in the United Kingdom, Shanghai in China, Melbourne in Australia, New York in the United States, Cape Town in South Africa, and so on. It is quite possible, even probable, that we have not been to all of these but have no doubt at all that they exist in the countries stated. The mass of cumulative evidence is so great that the knowledge is secure. We have read about them in a wide variety of literature; we read about them daily in our newspapers; we see pictures of them on television and cinema news programmes and features; we obtain vast reams of information about them on our computers; we hear stories about them from our friends and acquaintances, among whom one or other will have been to each. It is, of course, quite possible that the stories are not entirely accurate; it is quite possible that the pictures attributed to one city were actually taken in another or taken in a studio in a different country. We are probably right in being a little sceptical as to the accuracy of any media picture, but the overwhelming evidence of the existence of the city in question admits of no scepticism. Now move from large universally known cities, to smaller ones, and then to towns and obscure villages. We heard of a random shooting in June 2010, starting in the village of Lamplugh in Cumbria, with pictures of the location on the news and in the press. There was no reason to doubt that these were actually taken at the site, but one would not stake one's life on it! Compare the same day editions of a number of different newspapers reporting the same incidences as various television news: it is not at all unusual for there to be a wide divergence as to the facts as well as to opinions. Photographs claimed to be of this or that city, the accuracy of the reports, are matters on which we sometimes have to make a judgement as to whether we believe them, whether we know them to be true. The judgement tends to be trivial in the main, and secure. A photograph of the Matterhorn appeared in a broadsheet some years ago accurate except in the trivial circumstance of it being a mirror image.

In the realm of politics there are often matters on which two reasonable persons, sometimes close friends, have passionate and quite contrary beliefs, each thinking he *knows* which is the better policy, on ethical or economic grounds. If they are really reasonable, they will surely know that they do not know, and that it is so often an opinion based on little evidence.

There is perhaps the more serious issue of a very wide range of current scientific knowledge which informs much of our mores and following this, our behaviour. One example has already been mentioned: the definition of animal cruelty. We do not *know* that animals feel pain. We do know that actions which cause us pain result in similar neural responses in animals and we infer that animas feel as we do. There are theories which do not have an obvious effect on our behaviour, such as evolution, and ones such as of global warming which do. There is the morass of medical ethics and the difficulties caused when religion and science conflict, as in the state of the foetus. Very few of us have any first-hand experience in any of the beliefs underlying the relevant theories and no one in any but very few indeed.

No single person has studied more than a little of the evidence for the theory of evolution. Darwin compiled his monumental work following the observation over many years, of minor variations of isolated species, leading to his belief in their origin. There is no reason to doubt the authenticity of the work even though most of us have neither read it nor the very considerable literature which it has occasioned. Is it possible to say that we know that Darwin wrote it and that the illustrations are not fakes? It is certainly possible to read it and to come to other explanations: indeed, some of the evidence on which he built his theory is on the thin side. Add to his findings the thousands of other studies of the same nature and discounting a grand conspiracy it can be claimed that we know that the evidence is such. Add to this the very considerable palaeontological evidence, allied to increasingly coherent dating of clearly changing forms of both extinct and current species. This exhibits incremental changes, which can be seen over the centuries and the theory becomes secure. Once again, we must accept the evidence or have a totally unreasonable belief in a grand conspiracy. Finally, again taken on entirely reasonable trust, is the DNA evidence of thousands of species taken by thousands of scientists. That evolution has occurred cannot be reasonably doubted although the exact mechanics are still the subject of fierce debate. Nevertheless, an extraordinary number of Creationists, intelligent, educated, well informed citizens, all adherents of this or that religious creed as well as an even greater number of uneducated ill-informed members of the same creeds, do doubt it. Can either party claim to know where the truth lies? If either party relies only on religious belief, and allows this to override any evidence, then this is clearly not knowledge; if either party falsifies the evidence and relies on this to bolster belief it is deserving of contempt.

The dividing line between knowing and believing can be quite fine and is so in the case of evolution. The widely read person who has visited and studied the exhibits at one or more of the great Natural History Museums such as Beijing, Johannesburg, London or New York, or the smaller ones in virtually every country on earth, can claim to know that the theory is broadly true; the rest of us can really only claim a firm belief.

The theory (it is tempting to write 'fact') of evolution is easy to grasp, at least in its essentials. Some of the other major theories of the nineteenth and early twentieth centuries, which have so shaped our understanding of the universe, are a different kettle of fish. The first of these was the understanding of the laws of thermodynamics. To the law of the conservation of mass was added the law of the conservation of energy in any closed system. The ability to quantify the conversion of heat energy to mechanical work, allied to an understanding of the concept of entropy, placed theoretical limits on how much could be converted into useful work, and even these could never be fully reached in practice. The development of the steam engine, so much an essential part of the industrial revolution, was hugely informed by knowledge of the laws. Later they played a similar role in development of the internal combustion engine. Entropy is a sort of negative quantity, a measure of unavailable energy in any thermodynamic exchange. It always increases, whether in a closed system such as an individual heat engine or in the universe as a whole. The concept of entropy is also used in quantifying likely errors in the transmission of digital data. There is no doubt that a great deal is *known* in the field of thermodynamics but is also true that much is as yet – and perhaps always – that is not understood.

The formal treatment of thermodynamics began towards the middle of the nineteenth century. It was followed, in part by the same scientists, by the study of electromagnetism.[38] It is thought that there are four fundamental forces – more accurately, reactions – in the universe. Two of these, the strong force and the weak force, operate at sub-atomic particle level; gravity is the third and finally there is the

---

[38] On re-reading the manuscript it has become clear that much of the following paragraph has already been treated in Part 1 The Science: Mysteries of the Material World, but in what follows it will save the reader the tedium of referring back!

electromagnetic force which is responsible for the enormous range of electromagnetic fields. These are energy-carrying waves which permeate all space. The carrier is a fundamental wave-particle known as a photon and its wavelength varies from less than one billionth of a millimetre to tens of thousands of metres. The energy carried by each photon is inversely proportional to its wavelength. Gamma rays have the shortest and can affect physical biological structures, causing mutations in reproduction and cancers in the reproduced. Over-simplifying an extremely complex process, not only these, but every biological and every chemical interaction is occasioned by the absorption or emission of a photon by an electron. Light, visible to the human eye, has wavelengths of between four ten-thousandths and seven ten-thousandths of a millimetre. Radio waves vary from around one millimetre to over ten thousand metres and are imperceptible to humans. They only become able to be perceived by being intercepted by a suitable antenna, an aerial, converted into electrical impulses and transmitted to a radio which will convert theses to sound waves.

Late in the nineteenth and early in the twentieth two new sciences emerged: quantum physics and relativity. Newtonian physics, with the well-known laws of inertia, cause and effect, and conservation of momentum, which had been thought to govern every material body in the universe in a deterministic manner, were discovered not be true for objects at an atomic scale. Energy, like matter, was found to be particulate, and not infinitely divisible. Matter, at an atomic scale, was discovered to act in a probabilistic way rather than deterministic. Space and time which had been thought to be absolutes were shown in the Special Theory of Relativity to be interdependent, and energy and mass to be two forms of some underlying formless substance. The role of gravity in the nature of the space-time continuum was tackled, not entirely satisfactorily, in the General Theory.

There is as yet no unifying theory between quantum physics and relativity, and a great deal of controversy in the philosophical implications of each, but particularly the former. The theories become more and more abstruse, comprehensible only to advanced mathematicians; the experimental apparatus used to refine the theories become more sophisticated in their manufacture and cruder in their operation. Streams of elementary particles such as neutrons and protons are accelerated to nearly the speed of light in opposite directions by electric currents round two long tube-like structures with a circuit many kilometres long and made to crash into each other. The resulting splitting and behaviour of the particles is filmed and examined, and theories verified or falsified, or new ones proposed. The large hadron collider in Geneva is nearly thirty kilometres, buried deep underground. After a break of two years for refurbishment it is being switched on even as this is being typed[39], but the results will not be available for many months. The high energy collisions of the particles and their resulting splitting or joining are observed and current theories such as the nature of the plasma which formed the universe immediately after the Big Bang and the assumed existence of some elementary particles such as the Higgs Boson. The whole edifice of the present understanding of the nature of the universe depends on the verification of such notions. A little like the way the epicyclical motion of the planets, once the touchstone of scientific understanding, was shown to be more or less nonsensical, it is not wholly unimaginable that the present very obscure theories could be wholly wrong. It is much more likely that a completely revolutionary theory will add to current understanding rather than demolish it, much as quantum physics refined and built on Newtonian physics.

The verification or at least the non-falsification of a theory relies on experimental evidence and one of the problems is to separate meaningful data from random results, always present in any experiment. In this respect a standard deviation test is frequently used. A standard deviation, with the symbol sigma, σ, is a statistical measure of the likely proportions of divergences from expected results, with one sigma being such that 68% of the data are within it; two sigmas encompass 95% and three, 99.7%

"The number of sigmas measures how unlikely it is to get a certain experimental result as a matter of chance rather than due to a real effect."[40]

For example, the standard deviation of ten thousand tosses of a fair coin is fifty and one sigma gives a probability of only getting almost one third of the expected five thousand landing heads. A five-sigma deviation in a sample of ten million fair coins would result in a probability of only three not landing heads. Even this remote level of probability is not always sufficient in some particle physics' experiments before a declaration of near certainty of a theory is declared. It is only near certainty that is being claimed: no physical theory is certain. In manufacturing quality-control a six-sigma failure

[39] 22nd April 2022
[40] From a BBC website definition amended by David's blog in Understanding Uncertainty

rate is often used: between three and four defects per million tests is deemed acceptable.

That global temperatures are rising cannot be denied. The world-wide diminution of glaciers and pack ice are to a certain extent matters of personal observation. Any European over the age of sixty will know that ponds which froze annually in their childhood now do so rarely. So much is indisputable: we *know* that local warming is happening; there can be no doubt that global warming is happening although we rely on accounts in the media for this knowledge. What is not absolutely known are the causes, the progress and the likely consequences. Again, according to a huge number of media reports the evidence for some sort of cataclysm if stringent action is not taken very soon is becoming ever more surely based. Politics and the self-interests of meteorologists may well play some part in this, but the measures that are recommended are consonant with prudent and responsible behaviour and so should be followed even if the science is doubted.

While this is being written there is a war in Ukraine. The media reports on its causes, and to a much greater extent, on its progress, vary enormously in the Western press and the Russian. Those living in the West believe that their journalists are on the spot and are reporting accurately; Russians believe implicitly that their media are telling them the truth. On the one side there is (we believe) a free press; on the other (we are told) the press is rigidly controlled by the State. The freedom of our press is not as straightforward as might be believed. It is largely owned by media moguls who profess that they do not control their editors. Editors whose political views are contrary to those of the owners are very unlikely to be appointed so the question of overt control does not arise. There is an undeclared conspiracy between members of the appointment boards much in the same way as there is an often-unrecognised conspiracy, unrecognised even to the conspirators themselves, in nearly all professions, intent on preserving privilege. In trade bodies and trade unions it is more openly realised although here too there is often much self-delusion. Members confuse acting for the greater good of society, acting for that of their underlings or fellow members and acting for their own advantage. It is an unfortunate truth that the more respectable the newspaper or profession the more the conspiracy goes without recognition. Medicine and The Times of London are two good examples.

To return to the war in Ukraine. There are lurid claims of war crimes being currently committed by the Russian forces, which will be judged sooner or later by a not entirely impartial panel of judges. There are lurid Russian claims of atrocities committed over many years by the Ukrainian State against the pro-Russian citizens of the Donbas which are rarely, if ever, reported in our press and which are unlikely to be investigated but which might well be true. Even in events which we have seen on television, such as the declaration that there was no intention of invading Ukraine, we have to rely on the accuracy of the interpreters. It is unlikely, extremely unlikely, we think, that in that particular instance there was an entirely different narrative, but it is not impossible. In most human affairs we have sufficiently secure knowledge to be able to act as if it was absolute. Only when the occasion requires it should a buried grain of scepticism be allowed to surface.

Even that most scrupulous of moralists, Jane Austen, says in her heroine's words, towards the end of Pride and Prejudice, *Elizabeth was forced to give into a little falsehood here; for to acknowledge the substance of their conversation was impossible,* and towards the end of Emma, in her own words, *Seldom, very seldom, does complete truth belong to any human disclosure; seldom can it happen that something is not a little disguised, or a little mistaken . . .*

One of the determining influences on moral judgements is the evidence upon which they are based. Their dependence on knowledge gained from media is the subject for this section. Where they depend on authority and religious belief is discussed in Part 4.1 Morality.

Not only does the vast field of medical ethics rely on the accurate reporting of the evidence but also on the acceptance of this by the populace at large. For example, vaccines for the prevention of smallpox were widely promulgated throughout the 18th Century with no clearly documented evidence for their efficacy until at the very end of the century Edward Jenner carried out some well documented research. Following this, use of a vaccine became widespread in Europe and the Americas, but, unlike many European countries, did not become compulsory in the UK and Germany until the 1870's and never in some of the United States. Compulsion was always met with resistance and often by riots but governments stood firm and the widespread global use succeeded in eradicating the disease completely some hundred years later, when all known stocks of the vaccine were destroyed.

Children around a year old are routinely given the MMR vaccine preventing measles, mumps and rubella, with a second dose a few years later. The inoculation is not compulsory, but its efficacy and the dangers of the diseases are hammered home by health services and governments so that the great proportion of children all over the world receive it. There have been occasional cases of autism possibly associated with it or more probably (on what is this judgement based?) merely coincidental which have been the subject of what is claimed to be misinformation, and which have been followed by a decline in the vaccine being accepted and a large increase in measles. This results in a very much larger incidence of disease than the few numbers of autism, so if there is any government sanctioned misinformation this is very understandable. The possible very harmful effects of many policies on a small number of persons who would not otherwise have been harmed set against the avoidance of serious harm to the great majority is always a difficult judgement.

Similar behaviour is widespread in the fight against the current coronavirus (Covid-19). A number of vaccines have proved very effective in preventing serious illness and particularly death from the disease but none so far in containing it. The extent to which they have been used has been determined by the weight of propaganda used by national bodies and the public's acceptance of it as being trustworthy. The availability of the most efficacious ones and their worldwide dissemination has of course also played a crucial role but without the media endorsement they would not be used: by and large there has been no compulsion. As in the case of the MMR vaccine, a tiny proportion of possibly linked serious heart trouble causing far fewer deaths than would be the case if the population went without inoculation.

A similar effect has recently been reported in respect of "smart" motorways, where hard shoulders have been replaced by many monitoring cameras which ought to warn traffic if a vehicle using one has broken. Occasionally the system fails, and the unfortunate vehicle is impacted by a large truck travelling at speed, and its occupants killed. Public outcry over this horrible circumstance has been such that the project has been abandoned yet statistically the system has been shown to result in fewer deaths than would occur on a necessarily more crowded motorway. The trouble with statistics is that their accuracy, their truth as a fair summary of what is happening, is almost impossible to judge, even if one is very well informed. The old adage, lies, damned lies and statistics was never more true than today.[41] Quite apart from the difficulty of forming a sensible judgement on a set of data, there is the moral question of whether it is ethical to condemn a smaller random number of people to death rather than a much larger one. A different lot of fewer people would have perished.

There is a matter of grave importance on which scientific knowledge, ethical considerations and religious beliefs all play an important role: abortion. Infanticide is regarded with abhorrence throughout the 'civilised' world. This was apparently not the case with earlier well-developed civilisations and is still widely practised across the globe. This is only the subject of the present section in so far as it is known to be true and in this regard, there is a great deal of information readily available on the web. The morality of the practice will be discussed later. A foetus becomes an infant on birth and is thereafter treated as a human being, although there is some ambiguity of the status of stillborn children or even ones who die very shortly after birth. This is only relevant to this chapter in light of what is known about the development of the foetus. It is believed that between a third and a half[42] of all fertilised eggs are spontaneously aborted without the mother knowing she was pregnant or that it was happening. The only matter pertinent to this chapter is the accuracy of the statistic and the credence which can be placed on it. On this, the present author has no informed opinion. For the first few days after conception the embryo is microscopic; after four weeks it will be around one centimetre long and at nine weeks two and a half centimetres. At this stage its main limbs and organs will be apparent and from now on rather than being termed an embryo it becomes a foetus. The earliest that babies have survived early birth is twenty-two weeks; at twenty-four weeks they are generally viable and the normal full term is forty weeks. Based on these facts many countries have placed legal limits on the age at which an abortion may be done. Currently in most of Europe the limit is twelve weeks; in Germany it is twenty-two and in the UK twenty-four. In the USA the majority of the States have a limit of twenty-two weeks but at the extremes Texas has six weeks and Oregon, New Mexico and Colorado have no limits at all. The setting of a legal limit is largely a matter of public opinion, informed by understanding of

---

[41] Even the authorship of the adage itself is the subject of much dispute with, apparently, Mark Twain being responsible for an unfounded attribution to Disraeli.
[42] Wiki, quoting multiple sources.

the development of the foetus, by religious belief and dogma, and by feminism. The relative merits of these often-conflicting forces will be discussed later.

PART 4

RIGHT AND WRONG

1 Morality

All right then, I'll go to hell.[43]

There are a number of possible explanations for mankind's perception of right and wrong. A moral sense might be innate; it might be a result of evolution; it might be due to parental guidance, although in this case the question of where the parents got their guidance needs to be asked; it might come from religious teaching and this in turn might be due to revelation, real or imagined; it might, but not directly in the case of children, depend on a chain of reasoning. It is probably due to a number of these. Many moral norms, but not all, are in evidence across the globe and across the thirty or so centuries for which there are writings.

1.1   Conscience

I hope I haven't done wrong, but I couldn't, no I couldn't in conscience say she had done it when I was sure she hadn't, could I? Oh, dear, oh, dear![44]

Conscience is defined as *A moral sense of right, a sense of responsibility felt for private or public actions, motives, etc.; the faculty or principle that leads to the approval of right thought or action and condemnation of wrong.*[45]
Conscience ideally ought not to be some ill-defined feeling although all too often it is. We can act in conformity with our conscience or against it. In the case of the former our actions are personally virtuous and in the latter, vicious. The formation of our individual consciences

---

[43] Mark Twain's Huckleberry Finn, when his humanity overrides the rigid moral code he has been taught.
[44] Strong Poison. Dorothy Sayers, Chapter 3
[45] SOED II 4

are discussed below under the titles of Innate Moral Sense, Education (parental, school and self) and Responsibility.

1.1.1 Innate moral sense

The strongest argument in favour of an innate sense of what is right comes from an appeal to introspection. In nearly all situations we all know what our right course of action is. This is not the case where we are asked to judge others' actions, nor when we are theorising about morality. There are two difficulties with introspection: the first and greater is that I can only speak for myself; the second is that very often I do not know my own mind. Nevertheless (and here I speak for myself, in the first person – I can no other) even when I have been put in an impossible situation, where the consequences of doing what I considered to be right were dreadful, and I argued myself into an easier course of action, I knew what was right. But then, even if later reflection did not lessen my shame, might I have been mistaken and my action was in fact virtuous? This is unanswerable: we can neither judge ourselves nor others. There is a thought that if we were able to still ourselves into a condition of complete simplicity[46] we would *know* what is right. If not actually a delusion, the problem with this idea is that we can never be sure that the condition has been reached.

In both moral and material situations we are proud or ashamed or feel guilt when pride or shame or guilt are not in question. Beauty or deformity are obvious physical examples of matters which are often quite outside our control, yet we can, and frequently do, feel inordinate pride in the case of the first and embittering shame in the second. Guilt should only be felt when we are conscious of acting against our conscience. Feelings of guilt following actions which are entirely outside our control are commonplace: for example, the failure to live up to a wholly unreasonable expectation.

The self-attribution of wisdom in making decisions which are validated solely because of later unforeseen circumstances or events is very common, as is the opposite, the blaming of events which ought to have been foreseen for decisions which turn out badly. Bravery is not unfrequently merely the consequence of insensitivity and being afraid is not being cowardly. Indeed, bravery does not lie in absence of fear but the overcoming of it: in not allowing terror to interfere with action. Non-culpable ignorance is so often taken to be stupidity, both

---

[46] cf Eliot, in the final section of Little Gidding.

by the ignorant themselves, particularly children and more culpably by others. In these and countless other situations we have no worthwhile innate knowledge.

A more difficult argument against an innate sense of right and wrong lies in the near impossibility of discounting ideas which have been drummed into us from earliest infancy, predominantly by parents.

A second, and almost incontrovertible, argument in favour of innate knowledge at least in some aspects of morality comes from evolution. In common with most of the rest of the animal world we have evolved to protect and cherish our young. Even extreme dictators – Stalin and his 'little sparrow' daughter, Hitler and the children at his Alpine hideaway – there are numerous examples of genocidal maniacs or mass murderers behaving in a kindly way towards children, actions which cannot be dismissed as mere sentimentality.

There is a certain asymmetry in the notion of an innate ethical sense. There are virtues which appear to be common across the globe and over as many centuries as there has been writing. Kindliness, bravery, adherence to truth are cherished today in all civilisations. In particular in the early Latin poetry with which, to continue in the first person, I am cognisant, these virtues are very apparent, and doubtless in many others. On the other hand there is a ruthlessness, a glorification of savagery, in for example, parts of the Old Testament, in the Iliad, in Beowulf, which would be utterly condemned in most of the 'civilised' world today.

What then of the sentiment attributed to Huckleberry Finn by Mark Twain. He had been brought up to believe in the sanctity of the ownership of slaves and that to aid and abet their escape was a grave moral evil. When he had to decide what to do about Jim, a runaway slave whom he had befriended, he decided not to turn him in. This might have been an instance of an innate sense of right, contrary to the moral guidance he had been given, in which case it was an act of undisputed virtue. It might have been a selfish desire to protect a friend. The shielding of a criminal child by a doting parent is not virtuous. Huckleberry Finn, a fiction, was published twenty years after slavery was prohibited throughout the United States (in 1865 in the Thirteenth Amendment) but set some time before the Civil War which had started five years previously. The war ended in 1865 with the defeat of the Southern States in which slavery was still permitted.

In the end it probably comes down to a matter of faith rather than a reasoned analysis. To intrude personally again, my own belief, based on a somewhat doubtful faith and a far from clear introspection, is that if |I can still myself sufficiently then I *do* know what is a right course of action and that when I try to reason why it is so, I come to the same conclusion.

1.1.2   Education

Almost from birth children imbibe particular ethical and behavioural norms from their parents. These are the dominant influence in pre-school years and, to a greater or lesser extent, remain so for the remainder of their lives. The guidance might be serious, as in the teaching of selflessness, or trivial, in matters like table manners.

As children grow up some (but by no means all) of this guidance is questioned and, in what ought to be a lifelong endeavour, refined and deepened. Perhaps it is unfortunate that the age of childbearing occurs before this process is much advanced and parental teaching changes only slowly from generation to generation.

Sometimes, even more influential than parental indoctrination is that from primary and secondary education. In the case of religious schools, particularly if they are boarding, this can be the dominant agent in the formation of conscience, not necessarily moral conscience. A ninety-year-old was recently heard confessing that she still felt guilty if she used a safety pin in her underwear: her convent school had taught her that this was never done by a lady.

As far as this essay is concerned, the only concern is not in the nature of the guidance but the confidence which can be placed in any personal moral beliefs.

Core moral beliefs are to be found in a very wide variety of religious and philosophical writings. The earliest extant of these are the Ten Commandments, several hundred years before Christ. They are believed by Jews, Christians and Muslims to have been given by God to Moses and to have His authority. The first four are theistic and play little or no part in the moral teaching of the very many of a secular bent. The fourth, to keep Sundays holy and to refrain from all work has been watered down in all but the strictest of the Abrahamic faiths, and in most of Christianity even the prescription to go to church is widely ignored. The final six, filial piety, murder, adultery, thieving, libel and covetousness are universal virtues or vices. The writings of Confucius and Buddha, two or three hundred years later do not have such a concise list but like the Old and New Testaments are a collection of teachings.

Each religious sect has its own teaching of right and wrong and perverse interpretations have led to untold evils in many religious wars, and within each religion to unspeakably narrow and cruel behaviour. There are obvious perversions such as the Spanish Inquisition, the persecutions under Henry Vlll, Mary and Elizabeth, the rampages of Cromwell in Ireland and current Islamic terrorism. Christian moral teaching differs markedly in some respects from Muslim. Within Christianity, Catholics are told, dogmatically, that abortion is always a grave evil but its members use abortion clinics in almost the same proportion as non-Catholics; the same is true in the use of contraceptives, similarly proscribed. On the other hand, some Methodists teach, inter alia, "…if sex serves purposes beyond reproduction, then a woman has the legal right to both prevent and interrupt a pregnancy".

The fundamental difficulty is due to the fact that each religion believes that its creed is God-given: in Judaism via the Prophets whose revelations were then written down with scope for textual corruption. The teachings of Christ were recorded, many years after his death by the Evangelists and expounded by St Paul who never met Him. Paul's language in particular is often obscure and has proved a fertile ground for misunderstanding. The Roman Catholic Church claims infallibility in its reading of the text. In Islam the Quran is believed to have been revealed to Mohammed by God through the archangel Gabriel and to have been accurately recorded. The claims of infallibility of the Church, the Sola Scriptura doctrine of some Protestant sects, and the Quran have led and continue to lead to a dogmatic harshness resulting in confrontation, religious wars and terrorism.

Within Christianity it is a scandal of unimaginable proportions; within the Abrahamic religions it is a great misfortune for there is very much that they have in common.

1.1.3   Responsibility

The notion of responsibility cannot be separated from that of being held to account, both on the effect of an action to ourselves and on the effect externally on others or the environment.

Human responsibility rests on a belief that one action is good and that another is bad, and that we have some freedom in our choice to do this or that. The judgement of right and wrong is a moral one; the ability to act in accordance with the judgement is a matter of personal freedom. Without some genuine freedom (not just an illusion of it) there can be no responsibility. If belief that we do have some freedom and some responsibility for our deeds is universal, as it is, it is also the case that there are always some limitations, some mitigating circumstances which lessen our responsibility. It is true that there are many actions which we take to be free which are in fact merely reflex ones although their formation will probably have some moral implications.

There are two insurmountable problems in understanding the exercise of freedom. The first is the physical one of what part of our constitution is responsible for choosing to do this or that. There is no 'me' within my body or brain giving orders. Every action is the result of millions of neurons firing and emitting electrical impulses along axons and dendrites, and ultimately down nerves leading to muscle contractions and movement of limbs. Thoughts themselves share the same frenetic neural activity. It is possible (easy even) to imagine the sort of activity which occurs prior to any animal activity – a cat catching a mouse for example. The sight, sound and smell of a mouse will trigger a very definite response or series of responses in the cat's brain accompanied by an explosion of neural activity leading to the pouncing on the mouse and playing with it. Every nervous response, every single firing of a neuron is just a chemical reaction: there is no element of freedom at all. The same sort of procedure occurs in every human action. When a ball is seen hurtling towards one's head the person ducks. It is a reflex action, not a free one. When batters start facing fast bowling their instinct will be to avoid being hit, as this is painful. By practice and determination, they can school themselves not to get out of the way but take the ball and subsequent bruising on their bodies, until the obvious reflex action is negated. Here, there has been an element of freedom – unless the decisions to take their punishment were themselves reflex actions. It is quite possible to argue that no human act is in any way different from that of an animal, and indeed the argument is often made. The trouble with this is that no one believes it: we all know without any possibility of doubt that we do have some personal responsibility for our actions which would not be the case if we had no genuine freedom. What we do not know, is the extent to which we were free and responsible and where we were not. Introspection is useless in this context beyond making us aware of our inability to form any reliable self-judgement. Much that we feel ashamed of doing has no shame attached to it and much of which we think is virtuous has no virtue in it.

The effort of introspection ahead of any proposed action is very unrewarding. Try to be aware of what is going on in our head when deciding to go on reading this or pausing to go and make a cup of coffee. The act of putting off doing it is more or less discernible, perhaps because it is a continuing action lasting for a appreciable time. The decision to do so is virtually impossible to pin down. At some stage we just get up and make our way to the kitchen. No discernible decision has been made, and certainly no form of instruction from part of oneself to some other part. In retrospect, was the action free? One would like to think so. We are pretty sure that there was always the possibility of not pausing in our work. The sort of inner tussle which occurs when trying to get over some addiction, smoking, drinking, chocolate, whatever, is commonplace. While the resistance goes on some sort of positive determination is apparent; the moment of this determination ceasing is not so clear. It does seem to be a negative act rather than a positive determination to break the abstinence.

It is possible to imagine oneself standing before an all-knowing judge and being condemned for acts thought to have been altruistic shown to be merely being self-serving[47], or on the other hand praised for actions thought to be cowardly which are shown to be brave.

There is another problem. If we really do have some freedom, and it is not at base an illusion, then some actions have to be able to control the statistics of the quantum world. What would have happened in the natural course of events, and which is wholly subject to the laws of physics operating on every action and reaction from the Big Bang to the present day if some are to be bent to the will of a person. The problem of the mode of interaction between a spiritual substance and a material one is discussed in Part 5.1 The notion of ensoulment.

---

[47] Things ill done and done to others' harm
   Which once we took for exercise of virtue.   Eliot, Little Gidding

So, there it is. We know we have genuine freedom, not just an illusion of it, but cannot with certainty say of any action whether or not it was free, or to what extent it was free, and the problem of any possible mode of spirit and matter interaction remains.

2   Religions

So many religions: how can one possibly come to any judgement into which if any may be wholly or partly true? Our individual moral beliefs are so hugely influenced by our religious belief. One's standards of morality stem from parental influence; to an even greater extent so does one's religion. Paradoxically, whereas the former tend to last a lifetime, the early religious beliefs often change radically during teenage years. Even when this does not happen, having friends with other beliefs allied to becoming aware of the overwhelming scepticism of very many, probably the majority, of scientists leads to a diminution if not annihilation of a particular dogmatic creed. With maturity too comes a growing awareness that much of what is claimed in the name of religion is really political. Protestantism in Northern Ireland, the wearing of the burka in Europe, antisemitism the world over are all clear examples.

How many members of the Catholic Church would have been so had at least one of their parents not been Catholic? How many Protestants? How many Muslims? How many Buddhists or Hindus? Whereas it is possible that all are delusional to a greater or lesser extent it is not possible that more than one is true. Comparative Religion has a place here only in so far as it can be reasonably argued that some shared belief is right. Judaism, Christianity and Islam all believe in the one same God. It is possible that other religions do so too, even if not so explicitly. How then, as a minimum religious requirement, can it be argued that there is a God and His nature defined?

Plato in mystical language and his pupil Aristotle in logical, were the first to attempt proofs of the existence of a non-material God, the prime cause and sustainer of everything. As well as mystical language Plato used an ontological argument which was more clearly articulated around a millennium later by Anselm. A further two hundred years were to pass before Aquinas re-articulated Aristotle's logical arguments.

A brief example of each of these follows:

First, from the culmination of Diotima's discourse in Plato's Symposium:[48]

---

[48] Translated by Benjamin Jowett

Remember how in that communion only, beholding beauty with the eye of the mind, he will be enabled to bring forth, not images of beauty but realities (for he has hold of not of an image but of a reality), and bringing forth and nourishing true virtue to become the friend of God and be immortal, if mortal man may.

Secondly, Anselm's Ontological Argument:[49]
> *1.We can conceive nothing greater than God.*
> *2. It is greater to exist in reality than just as an idea.*
> *3. If God exists only as a thought, then a greater being can be conceived.*
> *4. This contradicts 1) above.*
> *5 Therefore, the greatest must exist in reality as well as in thought.*

And finally, Aquinas, following Aristotle, adduced his five proofs of the existence of God[50]:
> *1. The Argument from Motion: Our senses can perceive motion by seeing that things act on one another. Whatever moves is moved by something else. Consequently, there must be a First Mover that creates this chain reaction of motions. This is God. God sets all things in motion and gives them their potential.*
> *2. The Argument from Efficient Cause: Because nothing can cause itself, everything must have a cause or something that creates an effect on another thing. Without a first cause, there would be no others. Therefore, the First Cause is God*
> *3. The Argument from Necessary Being: Because objects in the world come into existence and pass out of it, it is possible for those objects to exist or not exist at any particular time. However, nothing can come from nothing. This means something must exist at all times. This is God.*
> *4. The Argument from Gradation: There are different degrees of goodness in different things. Following the "Great Chain of Being," which states there is a gradual increase in complexity, created objects move from unformed inorganic matter to biologically complex organisms. Therefore, there must be a being of the highest form of good. This perfect being is God*

---

[49] Proslogion, Chapter 2 – simplified version.
[50] Summa Theologica, prima pars, quaestio ll, articulus 3. Summary from open.library.okstate.edu, Philosophical Thought, Unit 2: Metaphysics 24

5. The Argument from Design: All things have an order or arrangement that leads them to a particular goal. Because the order of the universe cannot be the result of chance, design and purpose must be at work. This implies divine intelligence on the part of the designer. This is God.

In the following Articles the nature of God is expounded.

The fundamental objection to each of these proofs is the difficulty of providing an intelligible answer to the question of terminating infinite series by the bald assumption that there must be a Being outside the chain. The old adage, οὐδὲν ἐξ οὐδενός,[51] *ex nihilo nihil fit, nothing comes from nothing*, is not clearly articulated until a millennium later than the first exponent, Parmenides, and it was regarded as a false proposition by its proponents.

In a modern context the Big Bang poses the same question, and provokes the same controversy: if it is necessary for there to be a quantum field for the eruption to occur, where does the field come from?

Nothing approaching a universally accepted logical proof of any divinity has yet appeared.

There was a great flowering of attempts to prove God's existence in the 17th and 18th Centuries. In essence these are various versions of the mystical (Platonic), the Ontological (Anselm) and the Logical (Aristotle and Aquinas). They are neither more nor less convincing than the earlier attempts.

It is true, of course, that an individual's belief in a God or gods does not rest on any logical argument or demonstration. The dominant reason is parental indoctrination – the word is not being used pejoratively. Babies are taught about Jesus or Mohammed, the Buddha or Krishna, Thor or Odin and so on, at their mothers' breast. Parents have been taught by their parents from time immemorial. The possibility that the start of this chain was by divine revelation cannot be summarily dismissed, particularly from arguments that beg the question (in the sense of **petitio principii**).

It is now accepted wisdom that all life has evolved from a single organism; whether homo sapiens has evolved from a single couple is

---

[51] Parmenides, Ed Diels, Fragment 8; Lucretius De Rerum Natura 1 154

less clear. A large research study in 2018[52] came to no cut and dry conclusions: indeed, contradictory interpretations of the research itself have been published. It is far from impossible that the myth of Adam and Eve has a substantial basis in fact and that belief in God was a matter of direct revelation but equally it is perhaps more believable that it was not!

It is undeniable that most humans across the globe, today and through the centuries, believe in some supernatural being or beings, and wishful thinking is hardly a reasonable justification for this. At moments of great stress or when in the presence of some enormous aspect of nature, a storm at sea, an awe-inspiring snow-capped mountain, the immensity of the sky on a clear night free from light pollution – there can be few who have not had the experience of being in the presence of some huge mystery. Whilst this is not a vindication of a theistic belief it is a help. It is at least a counter to the powerful arguments brought forward by so many philosophers (not worth a great deal overall) and scientists (worth only a little more when straying from science into philosophy. And this present essay? God only knows.)

Quite apart from our first parents, if there be such, there are thousands of claims of a great variety of persons to have received revelations. Whereas most of these have little effect on societies, some by men who do not claim to be divine, are world changing. The Prophets of the Old Testament on Judaism; Saint Paul on Christianity and Christianity on the world (not always to the good); Mohammed and the foundation of Islam (and, for all that it is a perversion of the belief, much of today's terrorism) are just three examples. These three religions profess belief in the same God and conversion from one to the other is not so very unusual. Conversion from any of them to Hinduism is much rarer. Buddhism is another matter as it does not profess belief in any God but has its roots in Hinduism. It certainly has a large element of the supernatural in it and the language of mystics is almost identical in all religions. All the major religions too have their secular adherents who proclaim their virtues while denying any supernatural element.

The question of how it is possible to form a reasonable judgement of the truth of any particular creed is almost unanswerable. There is a huge and bewildering literature in which every nuance and aspect of every religion is propounded. There is no clear evidence that any of

---

[52] Stoeckle & Thaler in Gene Survey, Human Evolution 2018

90

the assertions are correct. Many of the assertions are contradictory and many based on misunderstandings. For example, Mohammed's condemnation of the Christian doctrine of a triune God rests on a wholly simplistic interpretation of the doctrine. Christian attempts to define the belief in the Trinity are so subtle as to be on the very edge of the limits of intelligibility (if not just a little beyond them): One God, all-powerful Father, maker of heaven and earth, of everything visible and invisible; one Lord Jesus Christ, the only begotten Son of God, begotten of the Father before all ages, Light from Light, true God from true God, begotten, not made, consubstantial with the Father....and the Holy Spirit, the Lord and giver of life, who proceeds from the Father, who with the Father and Son together is worshipped and glorified . . . .[53]

And then Christ, God and man, wholly God and wholly man. One God with three Persons; one of the Persons having two natures: it is not an easy doctrine.

There are truths which await outcomes if they are to be verified. The little red catechism of the Roman Catholic Church defines faith as being a supernatural gift of God by which we firmly believe in Him. It goes on to say that we must believe what God has revealed, and we know what he has revealed because it is the teaching of the Catholic Church. If this is true, it is not the obvious circular argument it appears to be, and no further discussion is needed, but how can anyone tell if it is true? There are times in the life of every religious person (of every person?) when there is no doubt at all that there is a Being outside the normal world of our experience; there are other times – and this is universal – when all is doubt and obscurity. Personal experience is not much help. There are devout persons of many different religions who sincerely profess a belief in a Divinity, in a startling similarity of terminology; there are billions of people, educated and uneducated, clever and stupid, who have a great variety of creeds with irreconcilable doctrines.

Belief in a one loving supernatural God, creator of the universe, is easier to justify than belief in any particular sect. Belief in more than one God – specifically one of good and one of evil – is not easy to dismiss out of hand. Indeed, experience of all the evident evils around us might be thought to add weight to the supposition. The theory that evil is merely a negation of good, in much the same way that dark is a negation of light, can be cogently argued but is not wholly

---

[53] From the Nicene Creed

convincing. In common language (never to be lightly disregarded) we speak of a thing being positively evil.

In this matter Occam's razor in its popular formulation, *Entia non multiplicanda sunt praeter necessitatem[54]*, *Entities are not to be multiplied except through necessity,* can be used to give comfort. The razor has had and continues to have an almost uncritical influence on scientific creeds, but there is no absolute logical necessity for it. It can be stated in many forms, such as *The simplest explanation is the right one.* One God is a simpler explanation for the creation of the universe than two! More seriously, a coherent philosophy can be argued for one all-powerful, good, God without any need to bring in an equally powerful evil one.

Whereas it is possible to make intelligible and powerful arguments for the existence of a single benevolent all-powerful God, the same is not true for any particular religion. With the exception of Buddhism all the major religions rely on revelation, and even the Buddha was well schooled in Hinduism and greatly influenced by it. Speaking of the Abrahamic religions, there is no simple answer as to why, in the matter of which is true, Christianity should be preferred to Judaism, or Islam to Christianity. Christianity rests on a belief in the divinity of Jesus Christ and on the astonishing *fact* of his resurrection. If he did not rise from the dead then, as St Paul says,[55] *Christians are the most unhappy of men.* If one is a Christian and looking for some sort of evidence to booster belief, there are several layers of evidence which are unsatisfactory. The gospel accounts of the resurrection are not as clear as could be wished and contain a number of inconsistencies; the accounts themselves are believed to have been written decades after the event and the earliest existing written fragments are two centuries later.

The text of the Quran is thought to be nearly contemporary and although there are internal contradictions in it, it has only one author and has neither the advantages nor disadvantages of having four. Advantages in so far as tradition has it that two (Matthew and John) were intimate witnesses, one (Mark) probably a witness of some of the events and Luke probably not a witness at all. This gives a good cross-check on the veracity of the accounts. The disadvantage of having four separate and largely individual accounts is that there will inevitably be minor discrepancies. The first three, the Synoptic

---

[54] Due to John Punch, 400 years after Occam's death (Wiki)
[55] I Corinthians 15, 19

Gospels, are remarkably similar in the events recorded and are thought to have influenced each other. John's Gospel is much more a personal recollection and quite different.

If there is an answer to the question already asked of how anyone can form a reasonable judgement of the truth of any particular creed it does not lie in the realms of logic or evidence but in the personal relationship of the believer and his God and God alone will be the guide in this respect.

Augustine poses the question in the opening of his Confessions as to how one can search for God if they[56] do not know whom they are searching for and comes to the conclusion that if we search, we will find and if we find we will praise, and perhaps this is the best answer.

---

[56] In today's horrible cant.

PART 5

AFTER DEATH: A SPECULATION

1 The notion of ensoulment.

If it is a spiritual soul that makes humans uniquely human, the moment of its infusion is of critical interest, but there is simply no way of knowing when this occurs. Some religions and cultures believe it is at the moment of conception but the spontaneous abortion of probably between a quarter and a half[57] of all fertilised ova within the first few days, and without the mother ever knowing she was pregnant, make the notion that the embryo is a person very difficult to believe, and the thought that it is a potential person does not have much meaning.

A strong case, based on all but universal human behaviour, can be made for it to be the moment of birth. The moment of the first extraordinary intake of breath would be sentimentally perfect. No civilisation now or ever has regarded miscarried embryos as deserving the respect of a funeral and the summary disposal of still-born babies is commonplace.

If the moment of birth is attractive in the light of universal funerary practice, the age of viability has a certain satisfying logic: not even the most radical of feminists sanctions infanticide but a significant proportion of women, perhaps a majority do not consider that their unborn babies are persons. It is left to the State to protect an embryo which might be viable, and nearly all States place time limits on abortion which vary from six to twenty-four weeks. These are informed by the earliest term at which a foetus has proved to be viable, without necessarily invoking the concept of ensoulment. This is unsatisfactory in that as medical knowledge and practice improve the period is shorter and shorter. Most Western States place limits of sixteen to twenty weeks. Islam allows abortion under certain specific circumstances at terms varying from six to sixteen weeks, and without limit if the life of the mother is at risk. All legislation has historically been overwhelming made by men, with little regard or understanding of the plight of women. Irrespective of the possibility that the embryo

---

[57] Wiki, quoting multiple sources.

is a person in its own right, and irrespective of any legislation, some women will seek abortion for a variety of agonising reasons. There is a compelling argument that it is better regulated in a safe environment than in a back street.

The Roman Catholic doctrine of the Immaculate Conception would appear to endorse the belief of ensoulment at the moment of conception although its current teaching does not deny the historical view that this does not happen immediately.[58] No further guidance is given.

A pragmatic moral argument used by Islamic scholars, that although abortion is always immoral, in some cases it is the lesser of two evils and therefore allowable if the mother's life is threatened.

1.1   Spirit-Matter Interaction

Some of what follows must be looked upon as speculative suggestions but ones which offer a simpler solution to the body-soul problem than the traditional matter-and-form answer put forward by Aquinas.

Apart from a statement of the obvious, that, as has already been said, if there is a God then interaction is a reality, there is very little that can be said. (If there is no God then the problem does not exist!)

There is still a great deal unknown about the nature of matter, and it is in this that the answer must lie. It is thought that there was a moment, some billions of years ago, when the universe was created in a void: no time, no space, nothing at all. The matter and energy consisted in huge numbers of disparate wave-particles which acted in a possibly determinate manner but more probably random. The particles would act in accordance with universal probabilities rooted in their very nature. This would result in their behaviour, in accordance with the statistical laws of modern physics, and would allow us to frame physical 'laws'. Why matter behaves in this way, why it had a proclivity to form ever more complex molecules, resulting in living creatures rather than just continuing in an expanding universe of separate particles is probably not a question that physics can answer. Physicists can uncover the laws but not the

---

[58] "It is true that in the Middle Ages, when the opinion was generally held that the spiritual soul was not present until after the first few weeks, a distinction was made in the evaluation of the sin and the gravity of penal sanctions." SCDF Declaration on Procured Abortion, 7.

reason for them. Consciousness, in the way that animals are conscious is part of the fundamental nature of matter: at no particular stage in evolution can it be said that this compound is conscious but that its immediate predecessor is not. If animals are conscious, in the human understanding of the word, this must be part of the as-yet unknown nature of matter. that it can be acted upon by a spiritual substance. The alternative is that living creatures are not truly conscious.

Human self-aware consciousness is a different matter. Here the spiritual substance, the soul, is part of the creature, able to act on its material to give it the ability to reason, free will and immortality. It is not the form of the body, in the Aristotelian sense, but part of it. The immediate form of the body is its genetic make-up, its DNA, able to be acted upon by its soul. (The ultimate form of DNA in the Aristotelian sense is rooted in its constituents, atoms, electrons and all the sub-atomic particles of which it is composed.) The soul has its own unique form, spiritual and imperishable, able to act on the body.

1.2 Spirit and memory

It is not a statement of the obvious that if there is a God, a spiritual substance can have memory. Whether God has memories depends on our understanding of His nature. If, as is generally thought, He is unchanging and unchangeable, then He knows all things in a single act, and memory is not an appropriate term. Memory is the recalling events which have happened in the past, or perhaps more generally the acting upon information stored in a creature's neural network. An injured swallow, nursed back to health, might have forgotten how to fly and the ability to do so be slow to return, but after a few faulty starts it does so, and then when the leaves begin to turn colour, or some other trigger, it remembers that it has to fly south and starts to gather on telegraph wires. These sorts of memory are mechanical: a computer could be programmed to act in the same manner – even to the extent of flying south! They are shared by man and beast.

In the physics of time and space, or perhaps more properly, in the philosophy of time and space, there is a controversial concept known as block time or the ugly term, *eternalism*. The 'commonsense' view of the universe as being three-dimensional with time a separate dimension, was shown by Einstein not to be feasible and a four-dimensional space-time is now universally accepted.

The concept of block time, or the block universe, was briefly mentioned in Part 1.2, Fundamentals of Space and Time.

The concept is simple: it is that the whole universe, past, present and future really exists as an unchanging entity. To prove that this is so by the theories of special and general relativity is by no means simple and so far as there is a consensus it appears not to accept the concept.

As an aside, the notion of time, like energy and mass, having a quantum, a minimum discrete existence, rather than being infinitely divisible, makes the understanding of block time slightly simpler.

There is no past or future with God – only an eternal present. In this present He knows everything that has happened, is happening and will happen until time comes to an end. In God, block time truly exists.

In human souls a limited block time exists: all that has happened and is happening; at the point of death this process is completed. During life on earth memory is exercised through a neural operation. After death it must be assumed that some sort of body will be needed to enable it to be accessed.

Some sort of analogy might be found in a flash drive. This can store data but needs to be plugged into a computer to enable this to be made manifest.

In a somewhat analogous way, the soul would store memories. It does not need a particular computer to release the data on a flash drive. Provided the hardware is compatible and a correct programme is used a huge variety of makes of computer would serve. With the human soul perhaps similarly, any human body would do but more likely it would need one with identical memory cells, which in turn would need an identical DNA – in fact not any human body but only one unique one.

1.3 Memory after death

After death the body disintegrates, and all neural components of memory are completely destroyed. The soul is spirit and imperishable and all the memories which are present in it at the moment of death remain but cannot be accessed without the soul's material body.

This is a dogmatic statement and one for which there is neither evidence nor any sensible suggestion of how it could even be possible to obtain any.

In the introductory note to this essay, it was assumed that sure knowledge was both possible and widely spread: it might be thought that the previous chapters have gone out of their way to counter this thought.

In developing a coherent theory of the nature and certitude of knowledge it is necessary to examine the possibility of interaction of the spiritual with the material. In Part 1, The Science, the notion of the solidity of objects, such as a billiard ball, was shown to be wholly inadequate if not actually illusory. Although there is much that is still unknown, we are reasonably sure that all material objects are composed, relative to us, of minute, sub-microscopic entities. These behave both as waves and particles which are themselves forms of some underlying formless substance which must exist as one or the other. They are interchangeable. A shimmering, largely empty world emerges, existing in a spacetime continuum replete with (composed of?) huge numbers of electromagnetic fields (light and radio, heat and high energy) and pervaded by the force-field gravity.

The mechanics of the interaction between immaterial and material objects, spirit and matter, has been the subject of much speculation and never satisfactorily answered. Consider the interaction between material and material: a billiard ball hitting another billiard ball. Assume, for the sake of argument, that the contact area is a thousandth of a millimetre square and a hundredth of one deep: a volume of one cubic micron. The number of atoms in this volume is measured in billions, some of which will react directly with corresponding ones on the other ball but the majority only with each other, diminishing with depth. Consider now the interaction of one molecule with another, and breaking this down, with atom on atom and finally electron on electron. Electrons are separated from atomic nuclei by greater comparative distances than the moon is to us, or a pea on the boundary of a football pitch to the ball in the centre. Electrons are both particles and waves. They can absorb and emit wave-particles of energy: photons which are the essence of chemical reactions. There is little meaning in the notion that they can touch each other. What appears to be a simple banging together of two small solid spherical objects turns out to be a hugely complex sub-atomic non-chemical interaction in which the meaning of *contact* is far from simple. It involves vibrational wave functions which are the subject of quantum mechanics, and which are still matters of considerable research.

Apparently solid objects can be regarded to some extent as an illusion dependant on the neural structure of the observer.

This is a difficult enough notion to come to any sort of understanding but the next topic is much more difficult, even if possible: the nature of non-material, spiritual, objects, assuming that such exist. Aquinas' prima materia is easy enough to understand. It assumed that all matter is conserved: it can neither be created nor annihilated but only change from one form to another. The example of a block of marble is often used, or a lump of plasticine: it is not possible for them to have no form; they can only exist in some form or other. Of course, plasticine itself has a form but there is a something underlying this, that which can be formed to make marble or butter, some underlying reality, prima materia. It is now known that mass and energy are linked by the famous formula $e = mc^2$. Under the right circumstances matter can be converted into energy and theoretically energy can be 'congealed' into matter. Mass and energy can be seen as two names for the same underlying, conserved, physical quantity: a modern prima materia. There is a theory that at base matter and energy are identical. They are not identical – just ask the citizens of Hiroshima – but different forms of the same underlying stuff, prima materia.

It is sometimes suggested that the interaction of spirit and matter takes place in the quantum world, the ever-active universe of unimaginable vibrating wave-particles. If there is a separate substantial spiritual world, this, of course, must be the case: it is so in every physical action or reaction, but is does not explain the mechanics of the operation at all.

We, mankind, have no senses able to experience pure spirit and are unable to form any imagination of it, or Him. We can theorise, can use inadequate language to try to come to some glimmer of understanding but this is all. We can practise the 'discernment of spirits', preached by the Jesuit founder, Ignatius Loyola in his Spiritual Exercises; we can remove ourselves as far as possible from all forms of human pleasure, can starve ourselves and meditate twenty hours a day for forty years as extreme ascetics of many religions have done from time immemorial and come to the state known as the specifically Buddhist notion of enlightenment or the Western one of the similar term, illumination. In this, according to a wide diversity of writings, after often a lifetime of searching in great obscurity of mind (the Dark Night of John of the Cross) one arrives at a profound sense of being at one with the universal Creator.

Is it possible to write anything sensible about the nature of a spiritual substance? The word sensible is unfortunate. Three forms of spirit are widely accepted to exist: God, angels and devils of various sorts, the human soul. Of these the first two are extra-spatial. That is, they do not exist in the space-time continuum that is the material universe although they can interact with it. While a person is alive the human soul ordinarily occupies the same space as the person. There is too much anecdotal evidence of phenomena such as telepathy, prophecy and dejà vue for them to be written off as nonsense. (Yet no sense is just what they are.) If these do exist they are examples either of the soul acting apart from the body, or it is just possible that there is an explanation to be found in the laws of physics? Theories of block time (or block universe or eternalism) provide a hint of how this might be. As has already been said, there are attempts to prove these theories but they are very difficult, very controversial and quite possible close to nonsensical. As that clearest of physicists, Richard Feynman, is often quoted as saying, *I think I can safely say that nobody understands quantum mechanics*[59]. The same can equally said of block time.

Belief that we, homo sapiens, have souls is almost universal – it is only in the last two or three hundred years that a scientific (as opposed to a philosophic) disbelief has become part of the accepted wisdom of the West. Those religions which believe in reincarnation identify soul with life: all living creatures have imperishable souls. In the Abrahamic religions the soul is, in Aristotelian terminology, the form of the person: it is that which makes the body not only a person, but specifically you or me.

It is immortal, imperishable, and survives after death either alone or, for resurrectionists, united to another, imperishable, glorious, body; for those who believe in reincarnation if might be a miserable or a glorious body.

Memory is a material function common to all living creatures and, it can be argued, to inanimate objects too, at least by analogy. A stone 'remembers' that it has to roll downhill! A swallow remembers how to fly; moreover, it remembers to fly away in autumn from wherever it has summered. These memories are locked into part of the creatures' central nervous systems. As yet there is very little understanding of the mechanics of this beyond some localisation of areas of the brain which are responsible for it. Theories are beginning to emerge, but only beginning.

---

[59] This quotation has many attributions but the wording above is possibly apocryphal.

100

As far as human memories are concerned there is no reason to think that these differ in any way from any animal memory. Memories are largely dormant and are stored in the neural network that is the brain. They can emerge into our consciousness either spontaneously or deliberately, but this latter is often elusive: we only have a very limited ability to remember what we want to remember. (Would that we could forget what we want to forget.) When a memory does emerge into consciousness it becomes a quale. The word *quale* (singular of qualia) in the sense used here does not appear in the SOED[60] although it does in the Wikipedia article on qualia. Qualia are things of which a person is subjectively self-aware. When the headache caused by trying to define the term is forgotten for a while, even though the neurological symptoms are still present, it ceases to be a quale. Memories are only qualia when they are summoned – or spontaneously arise – into our consciousness. Dreams are not strictly speaking qualia but the memories of them are. The nexus between qualia and non-qualia is one of which there is as-yet no understanding at all. There is much fierce debate and many theories, of which the commonest is that given sufficient neural (or electronic) complexity in any living creature, acts of qualia will emerge. The evidence for this in the case of computers is absolutely nil: for all the immense advances over the last fifty years they remain wholly intelligible machines with no sign of any evolution whatsoever in this respect. The position with the animal kingdom is not so clear. That animals are conscious is abundantly apparent, but it is difficult to imagine that a sheep has any trace of self-awareness as it spends its short life nibbling grass in the cold damp of the Highlands, and there is certainly no evidence for it. As New Forest ponies spend theirs similarly, similar difficulties arise. When it comes to a racehorse, or a much-loved hack the imagination is not so difficult but human behaviour to all animals, including pets, belies any belief that we truly believe that they possess subjective self-consciousness.

Qualia are unlike any other phenomena in the universe. That they are uniquely human is accepted by the vast majority of the race, including the scientifically educated. This of course is not a valid argument for the veracity of the statement.

There is no discernible neurological difference between a quale and any other conscious act, between an animal and a human conscious act. Wherein, then, lies the difference? All human conscious actions

---

[60] The New Shorter Oxford English Dictionary 1993

are qualia; no animal ones are. There is a huge quantity of matter dormant in our memories. Some of it is easy to recall, some (like the scent of madeleines) tantalisingly unrecapturable, some emerging only in dreams, decades after the event. They only become qualia upon recall. Whether dreams themselves are qualia is perhaps questionable: they are subjective but not fully conscious.

The greatest problem in belief in an afterlife is what happens to our memories. If, as seems unquestionable, memories entail the short- or long-term alteration of some brain cells where they are physically stored. These are more or less accessible, and their destruction is a necessary consequence of death's corruption. We can consider the human soul in Aristotelian terms as the form of the body, but unlike the form of a horse it is held to be both incorporeal and subsistent: that is, it can exist separated from that of which it is the form. This is more or less intelligible but even if true it does not take us much further in understanding of the nature of an immortal soul.

There are two distinct theories of the bond between body and soul: unitary and dualistic. The first uses the theory of matter and form first proposed by Aristotle and a thousand years later refined by Aquinas but used in a somewhat idiosyncratic manner here.

What is meant in this essay by the term *form* can be illustrated by the following. The form of a horse is contained in its genetic material, DNA. It is more deeply imbedded in the substance of the animal than the accidents of colour and size and such like appearances, and is shared by all members of the species. The form of the animal is here defined as being its genome, the entire set of DNA instructions found in its every cell.[61]

A horse's genome is what makes it a horse rather than a dog; a human's genome is what makes him (or her) a human rather than a gorilla. No chemical analysis will detect anything unusual in the genome: only a slightly different ordering of the nucleotides, adenine, cytosine, thymine and guanine (A, C, T or G) in the chains which are the genes. They are fairly simple molecules of carbon, hydrogen and nitrogen – the chemical formula od adenine is $C_5H_5N_5$.

In every cell in every human there are in excess of three billion nucleotides split among some twenty thousand genes which are

---

[61] Of course, in strict Aristotelian terms, the form of the genome is its structure, and the molecules of which it is composed are its matter: the form here is itself both matter and form!

contained in twenty-three pairs of chromosomes. The great majority of the pairings is common to all animals; the very great majority is common to all mammals; all but a small number are common to each human and each individual has a unique one.

If there is an immaterial element, a spiritual soul in man, and it is not to be found in the organization of his chromosomes, where is it? Either it is a separate substance, or substances located somewhere in the body, or a separate substance spread throughout the whole body. For the former various possible locations have been suggested from time immemorial and in many cultures. The favourite places are the heart and liver for a soul responsible for feeling and the brain for one giving man the ability to think. The first modern articulate description was Descartes' proposal that it was to be found in the pineal gland. Better understanding of the function of the gland has resulted in this theory being totally discarded. But of course, a dualistic theory does not need any particular location: a spiritual substance does not have spatial dimensions. The soul is not confined in any particular location but is wholly present throughout the whole body and wholly present in every part of the body. This thought was very elegantly suggested by Augustine,[62] without resorting to Aristotelian terminology. In terms of matter and form, the matter in a human body (in any body) is the prima materia underlying the atoms and molecules, the electrons and photons of which it is composed and its form, its activating principle is its subsistent spiritual soul. By subsistent is meant that it can have an independent existence, unlike the forms of everything else. According to this vein of thought, every single particle in a human body has its fundamental substance changed by the soul, its form. It is a beguiling theory, offering a tantalising possibility of some deep understanding.

The trouble with this theory is that even though it offers a glimpse into an unfathomable mystery, it is not far from being incomprehensible, so an alternative theory will be explored. This is that body and soul are separate substances, and in this respect two

---

[62] Cf De Trinitate Vl 6: Nam ideo simplicior est [anima] corpore, quia non mole diffunditur per spatium loci, sed in unoquoque corpore, et in tota tota est, et in qualibet eius parte tota est... (For therefore the soul is simpler than the body because it is not extended in mass through a particular place but in each individual body it is both wholly present in the whole and wholly present in any given part.)

problem areas will be examined: when they are first joined (the moment of ensoulment which has already been discussed) and the nature of the re-unification: abortion and resurrection of the body.

2 The Necessity of the Resurrection of the Body.

The argument in a nutshell:

> At some stage in a human's life the body receives a spiritual soul and becomes a person.
>
> If God, pure spirit, can interact with matter, so can the human soul.
>
> If God, pure spirit can know everything in a single act, so a human soul can know all its limited knowledge: can have all its intellectual memories preserved.
>
> Man gains all his knowledge through his senses and his memories are stored by some modification of his brain cells, in exactly the same way as all animals.
>
> The soul preserves memories of everything which the material memory has recorded, in a single act of memory, but it needs a human body if it is to be accessed.
>
> After death the physical memory is totally destroyed; the soul is imperishable.
>
> For this reason, the resurrection of the body is necessary. Without it this memory cannot be exercised, and without memory I would not be me. There is no point in somebody else going to heaven for my virtues!
>
> Not any body but my unique one is needed for the memories to be accessed.
>
> The uniqueness of my body is determined by my unique DNA.
>
> Therefore, the resurrected body will have my DNA, be the same body and be me!

These thoughts will now be teased out a little, taking each of the above in turn.

Well over half of the world's population profess a belief in the continued existence of the soul after death.[63] In Hindu doctrine this is accompanied with reincarnation into another animal form; within Christianity and Islam, resurrection of the body occurs sometime after death. There are at least two difficulties in trying to understand the nature of this. First in general: how the relationship of any spiritual substance to any material particle would then apply to a human spiritual soul's relation to the body and secondly what happens when the body disintegrates. Each of these favours a dualistic solution. This at least has the merit of being a little more comprehensible even if the problems of how and what remain.

How? Here we run into one of the basic difficulties which has already been discussed above: the mode of interaction of spirit with matter. Imagination fails and there is not much more that can be said, than if there is such a thing as spirit – if there is a God – then interaction is clearly possible, but we have no means of understanding it. This is far from satisfactory but perhaps should be used to try to get some understanding of the nature of a spiritual substance.

It is not all that unusual for human creators – artists, composers, writers and the like – to claim that they saw the whole of a work in an instant of inspiration. It was long believed of Mozart that a whole symphony was in his mind as a single act and that all he had to do was to write it down:

> *"Provided I am not disturbed, my subject enlarges itself, becomes methodized and defined, and the whole, though it be long, stands almost finished and complete in my mind, so that I can survey it, like a fine picture or a beautiful statue, at a glance.* **Nor do I hear in my imagination the parts successively, but I hear them, as it were, all at once.** *"*[64]

Although there appears to be considerable doubt about the authenticity of the letter in which this quotation appears, the substance of it is so believable that until recently it has not been questioned. In our recollections of music or poetry something very similar frequently occurs. There is no 'I' within me surveying the contents of my memory cells, and in this

---

[63] The old proverb, Vox populi vox Dei could perhaps be adapted to Vox populi vox veritatis.

[64] Often quoted from a letter supposedly written by Mozart.

experience there is perhaps a glimpse of the power of the soul to hold disparate memories simultaneously.

Let the majority of readers, if they will, dismiss this with a sneer; one or two will treat it more compassionately.

The final (final? Good God!) problem is what happens on death, and here too a dualist theory is much more straightforward than the matter-and form theory. The question of interaction now disappears. The question of memory remains, but a dualism does present a possible solution. Most of human memory is exactly of the same nature as that of any other living creature, involving the physical alteration of some brain cells which will someday be better understood than at present. It is only when they become qualia that their nature is very different. An act of self-conscious awareness is not physical. No chemical, neurochemical (chemico-neurological) or any other physical analysis will discover any specific alteration in any neurone. Nor, if it is a spiritual phenomenon would this be expected. This is not to say that neurones have no part in any qualia: of course they do. Without them there would be no qualia, no human thoughts, no insights. We need our complex brain functions to be able to have them at all. The soul is a spiritual entity. It has an intellectual capacity which is lacking in animals but no means of exercising this without a physical body. Now, the critical, hardly answerable, question is whether this spiritual soul can store memories.

As an aside, the current Catholic dogma of the resurrection of the body is that it is the same body, but what is meant by this? Our bodies are in continual change, not like a river flowing between largely unmoving banks and receiving its identity through this, but in a continual living replacement of identical or near identical elements. Every atom, even the ones in teeth and bones, one reads, is replaced every seven years or sooner. It is, one hopes, no part of Catholic dogma that the resurrected body is the one at the moment of death. If not at the moment of death, when? There is no single suitable moment in anybody's life (think of the child run over at the age of two, or the long sufferer from a variety of cancers dying at last shrivelled and twisted, blind, deaf and dumb.)

Paul's teaching in 1 Corinthians 15 is easier to come to terms with:
But someone will ask, "How are the dead raised? With what kind of body will they come?" How foolish! What you sow does not come to life unless it dies. When you sow, you do not plant the body that will

be, but just a seed, perhaps of wheat or of something else. But God gives it a body as he has determined, and to each kind of seed he gives its own body. Not all flesh is the same: People have one kind of flesh, animals have another, birds another and fish another. There are also heavenly bodies and there are earthly bodies; but the splendour of the heavenly bodies is one kind, and the splendour of the earthly bodies is another. The sun has one kind of splendour, the moon another and the stars another; and star differs from star in splendour. So will it be with the resurrection of the dead. The body that is sown is perishable, it is raised imperishable; it is sown in dishonour, it is raised in glory; it is sown in weakness, it is raised in power; it is sown a natural body, it is raised a spiritual body. If there is a natural body, there is also a spiritual body So it is written: "The first man Adam became a living being; the last Adam, a life-giving spirit. The spiritual did not come first, but the natural, and after that the spiritual. The first man was of the dust of the earth; the second man is of heaven. As was the earthly man, so are those who are of the earth; and as is the heavenly man, so also are those who are of heaven. And just as we have borne the image of the earthly man, so shall we[g] bear the image of the heavenly man. I declare to you, brothers and sisters, that flesh and blood cannot inherit the kingdom of God, nor does the perishable inherit the imperishable. Listen, I tell you a mystery: We will not all sleep, but we will all be changed, in a flash, in the twinkling of an eye, at the last trumpet. For the trumpet will sound, the dead will be raised imperishable, and we will be changed. For the perishable must clothe itself with the imperishable, and the mortal with immortality. When the perishable has been clothed with the imperishable, and the mortal with immortality, then the saying that is written will come true: "Death has been swallowed up in victory. O death, where, is your victory O death, where, is your sting?"

This is not quoted in justification of the present thesis but fits in well with it!

Not any human body but your own unique one, suitably resurrected, presumably by giving it the same DNA. It is not difficult to attach some imagination to this thought even if it is very woolly. It is virtually impossible (completely impossible?) to form any imagination of the way that a spirit with no spatial definition and no parts, would be able to store a vast number of separate memories. If

107

one believes this to be possible to God, as any theist must believe, then it must also be possible to other spiritual bodies.

The human soul is responsible for making material consciousness reflective, self-aware consciousness, qualia, which are non-material. They have their origin in the body and need it to operate. The soul on its own has no means of action: in itself, it is not conscious. It is in the same sort of state as we are when asleep and dreamless. This is the case for belief in the resurrection of the body: it is necessary if the soul is once again going to be able to act.

So much for the necessity. Can anything sensible be said about the nature of the resurrected body. Even for the barefaced speculation which underlies all this section, the answer as far as the present author is concerned is emphatically in the negative. This could well be a defect of imagination, or it might be that the imaginings are ruled out on the criterion of what is sensible. He has no sensible thoughts, so his advice must be, 'Wait and see!'

Will we know husbands, wives and children, friends and enemies? As there can be no worthwhile answer to these questions, they are better unasked!

Wait and see.

PART 6

CONCLUSION: SOME CERTAIN KNOWLEDGE – A COMFORTING SPECULATION

We all lead our lives unhampered by arcane philosophy, sure in the knowledge that our senses are feeding us true information. (This is not in any way at all to denigrate philosophers or the study of philosophy.)

The tree that we see[65] is a beautiful solid many-branched wonder which exists as we see it, independent of us. It is just as we see it, feels as we feel it, smells as we smell it. The extraordinarily complex nervous system by which we sense it corresponds to the extraordinarily complex composition of the object sensed, be it a tree or a ball bearing. Both are the work of the Creator. It is the work of the scientist to unravel, as far as possible, both these strands. It is emphatically not his business to theorise about the ultimate reality of what he is investigating. This is properly the province of philosophers and theologians – even amateur ones.

Human beings have evolved equipped with the sensory apparatus capable of becoming aware of a material universe fashioned to that end. When we see the sun rising in the East this is what is really happening. At the most basic level, the interplay of gravity and photon emission are the means by which this reality is accomplished: they are of course real, but subordinate realities.

This is a theory; it assumes the existence of an omniscient, omnipotent God. Its enunciation is the whole purpose of this work, but is there any evidence for it, anything which can withstand the scepticism, not to say derision, which it will attract?

There is a sort of negative evidence in that it does not lead to the contradictory conclusions in every materialistic theory. The impossibility of finding any sureness in in any of our various ways of

---

[65] in the quad or wherever

knowing has been laboured in the foregoing chapters. The means by which we receive sense impressions, which is the only way we can interact with the world around us, are extraordinarily complex and only partly understood. Indeed, in the case of qualia, acts of which we have a self-conscious awareness, there is no understanding at all of how this happens.

Objects which appear to be quite simple, such as a ball bearing, are hugely complex. Far from being solid, they are largely empty space, composed of molecules and atoms, which in turn have rather mysterious nuclei. These are a tightly packed group of neutrons and electrically charged protons, which in turn are both composed of quarks which have no independent existence. Each nucleus is surrounded by clouds of electrons which are both waves and particles and which can absorb or emit fundamental units of energy, photons, which are also wave-particles. There is still much that is not yet understood about this structure. Space is neither nothing nor infinite. It is permeated by gravitational and electromagnetic waves, of which two examples are visible light, which we can sense, and radio waves, which we cannot. It is possible that these, and perhaps other, waves actually constitute space.

Time is almost beyond understanding. The notion that the entire universe, its past and present and future, have an equal reality and that we, conscious creatures, only become aware of a little part of it, little by little, has been briefly discussed in the concept of block time.

Knowledge of the fundamental constitution of the universe is still very far from complete, and some or all of the current theories might be partly or wholly wrong. It is thought that less than one twentieth of the universe is available for our observation. The other nineteen-twentieths are in the form of dark matter and dark energy. They are called dark because they do not interact with any known electromagnetic field: they are impossible to detect except by hypothesis to make the sense of 'ordinary' physical theory.

This is not in the least to decry scientific research, to denigrate in any way the continuing efforts by thousands of dedicated or dilettante people to come to a better and better understanding of the wonders of the universe, a process which will never come to an end. The idea floated in the opening paragraph of this chapter, that the world as sensed is the ultimate reality is beguiling, and does bear considering. Analogies can be found in a Mozart symphony or a film. The symphony has many modes of existence, It is first created as a thought in the composer's head, inaccessible to anyone but Mozart. It is written down in musical notation on paper and contains full or nearly full instructions on how it is to be performed. Later it will be printed in hundreds of copies or made available online. Violinists, clarinettists and the rest of the orchestra will be given their parts and under a conductor will play the symphony and it will become what it was created to be. There cannot be much dispute that it is the played symphony that is the ultimate reality in this instance: the rest are like bits in a watch. The Jupiter can be recorded on a compact disc and played on a computer. Not a live experience perhaps but still surely the Jupiter. What about it when it is being played in an empty room? Is it still the Jupiter or does it need a human ear to turn it into sublime sound? Rather than get into a metaphysical muddle is to assume that it **is** the symphony and human beings are equipped to hear it. The tree does exist in the quad when there is no one to see it. The symphony does exist in the room when it is playing even if there is no one to listen to it.

The second analogy is a film which once again can be stored in a variety of media and can be shown in a variety of media – in a cinema, on a television set, on a computer screen. The shown film is the reality. It needed all the paraphernalia of script writers, of sets and actors, producers and sound engineers, camera men and their cameras, to become a series of images on a strip of cellulose or stored in digital form on some computer memory. It is clearly a film, just as its creator intended, when seen by an audience. Is it still a film when played to an empty cinema?

These analogies were chosen because they are both examples of the products of a creator.

If there is a Creator and the universe is His creation, then the problems of ultimate reality disappear. The world is as we sense it. The sun really does rise in the east, traverse the sky and set in the west. It really does: we see it every day and eyes do not deceive us. There is no difference between thinking that the reality is a system with planets going round the sun, giving their inhabitants the illusion that it is the sun which is moving and not themselves, and that a ball bearing is not a small shiny solid metal sphere but really an extraordinary largely empty extremely complex construction of unimaginable vibrating wave-particles. Why are we content to think that the ball bearing really is a little sold ball and be thought stupidly ignorant if we think that the sun really does rise in the morning?

There are at least three theories of reality.

6.1 Materialism

Nothing outside the material universe exists. There are no spiritual substances, no souls, no angels, no God. What things are, the nature of matter and energy, space and time, of human psychology and qualia are all not only the province of scientific research but will in the fullness of time be fully understood.

This is unsatisfactory as the evidence is that not only have theories throughout the whole of history proved wholly or partly wrong (and there is no reason to think otherwise now) but also the more that is uncovered in our understanding of the fundamental nature of matter, the more we become aware that there is so much more that we do not yet understand. It is rather like a Mandelbrot Set: every branch gives rise to similar small branches, and so ad infinitum.

The nexus between consciousness and self-reflective consciousness is not only as yet unknown but there is a strong argument that it is outside the scope of any physical research. And will therefore never be understood in purely material terms.

There is also no remotely satisfactory theory of the emergence of the universe from a void, an absolute void, devoid of any electromagnetic, gravitational or any other field whatsoever. Since it is clearly impossible for anyone to make a void in a laboratory it would never be possible to test any theory. Putting theories to the proof is the absolute bedrock of science.   The notion of the universe having always existed, while not susceptible to disproof, is profoundly unsatisfactory.

One constantly hears the argument that the assumption that there is a God who created the universe merely puts the problem back one step: who created God? The classical answer to this is that infinite cause-and-effect chains are repugnant to reason and that the only answer is an uncaused cause. Whether the one is more repugnant than the other is unarguable: in the end it comes down to personal belief, or perhaps personal inclination. In the same way that it is beyond the imagination of some as to how purely material objects could possibly evolve into self-conscious entities, others reserve judgement, and yet others firmly believe that not only is it possible but that it has unquestionably happened and that one day the *how* will be understood.

6.2 Idealism

The theory is that objects only become themselves when they are consciously observed by a human. Until this, they exist in some sort of potential, which is the province of science. They only become real when they enter consciousness. For example, the music from a radio playing in an empty room is only a system of pressure waves in the air: it become music when the waves act upon an ear, are transformed to electrical impulses which travel along the auditory nerve to the brain and enter the hearer's consciousness.

Different types of reality exist in the passage from object to perception and there is perhaps little to be gained in trying to classify them in some sort of table of lesser or greater. It is easy to read some Idealist philosophers, such as Bishop Berkeley, and interpret them as saying that the material universe simply does not exist at all but they are very much more nuanced than this.

It is difficult to understand the exact intention of Berkeley, but he was certainly not saying that material objects had no existence at all When it comes to not being able to understand something there are two possibilities. Either it is due to a deficiency of one's intelligence or because the idea itself is wrong. It is impossible to understand something where there is nothing to be understood![66]

6.3 A theistic solution

The third theory of reality is as stated in the opening paragraph of this chapter: that the things which we sense are real, are the work of a Creator and a theistic solution to the problem of what is real is is neatly expressed in two limericks criticizing Berkeley. The first is certainly, and the second probably, by Ronald Knox:

> *There was a young man who said "God*
> *Must find it exceedingly odd*
> *To think that the tree*

---

[66] As a personal footnote, I use this argument to justify to myself many philosophical opinions which I find difficult or impossible to understand. Kant's Transcendental Idealism is a good example.

Should continue to be
When there's no one about in the quad."

"Dear Sir: Your astonishment's odd;
I am always about in the quad.
And that's why the tree
Will continue to be
Since observed by, Yours faithfully, God."

The alternative is that the music really exists in the empty room and is available to be listened to by anyone who enters the room. Is the music a glorious sound or a lot of pressure waves? Then, are pressure waves themselves simple entities? Is it necessary to reduce everything to their basic constituents, the mysterious particle-waves if the search for reality is to be ended? If the former, when does it change from pressure waves to music? Is the suggestion sustainable that the music in the empty room really is music? It has the clear merit of being much simpler, but also more difficult to understand.

One thing is quite clear: objects sensed are not altered by being sensed. The tree is not affected by our seeing it; the music in the empty room is exactly the same as the music in a room filled with enraptured listeners. The music in the empty room is music, ready to be heard with anyone within hearing distance. It is the old argument about a tree falling in the forest with no one to hear it. Does it make a noise? One cannot deny that it causes a precisely defined pattern of air compressions but that these are the noise of a falling tree, the actual noise, which any well-functioning ear could hear is not so straightforward. Is there any difference between this reality, and the sensation of hearing? Are the extraordinarily complex operations which occur within our central nervous systems when we hear a sound (particularly one which makes us jump or tap our feet) the reality or merely the means by which we become aware of the reality? The noise which we hear is the reality: this is obvious. It is the assertion that our senses are the instruments by which we perceive reality which is being argued here.
There is no evidence for this theory, nor can there be: it is unarguable. It is a beguiling point of view and one that leads to a commonsense view of the universe shared by the vast majority of the human race. It is irrelevant whether there are any listeners in the forest. Of course,

the tree makes a noise when it falls. There might be some evidence of this in a permanent trace of leaf movement due to the sound's shockwave, or there might not be any. That there will have been a shockwave is not in dispute.

Of course the sun rises in the morning and sets in the evening. There is no difference in principle between the path of the sun and the noise of the falling tree. The physical cause of the noise is a pattern of air compressions. The physics behind an air compression is both extraordinarily complex and part of an infinite chain; the motion of the sun round the earth is similarly complex and part of a number of unending chains of cause and effect, the answers to the questions, Why? Why? Why?, questions at any moment of time extending to the furthest frontiers of space and questions stretching back through time, right to the Big Bang, if there was such a thing.

Perhaps the least that can be said about this way of looking at the universe is that it is a serious argument in favour of common sense. The sun does rise and set; the tree falls with a resounding crash.

It might be that this is merely a matter of semantics, but what follows is certainly not. It is a statement justifiable only by faith.

Whatever is susceptible to human senses is their ultimate reality.

We were created by God and have evolved to be what we are today; we were created, have evolved, bodies as part of the general evolution, souls unique to each person with the ability to become aware of the of the material universe.

What we see, we really see: the sun setting in the glory of an autumn evening, newly unfurled beech leaves in the splendour of a spring morning.

The reality is out there and provided our senses are in good order we can become aware of it. If we are hard of hearing the beauty of the thrush's ecstasy is diminished only for us.

The ultimate answer for all Christians lies in Christ, the Word of God. He was with God and is God.[67]

When the Father sees the material world, brooded over by the Spirit, He sees it through the eyes of the Son.

---

[67] John 1.1

To sum up[68]:

Apart from purely 'animal' knowledge, such as how to walk, and possibly some innate knowledge such as a discernment of good and evil, everything we know comes by way of our five senses.

This entails an extraordinarily complex activity of huge numbers of sub-atomic particles involved in dynamic neural responses between epidermis, eyes, ears, nose and tongue, and brain and within the brain itself.

Where this knowledge is direct, as in seeing and hearing,

it can, except in rare circumstances, be relied upon: what we see is as we see it.

We have no non-sensual knowledge of another person.

When the knowledge is indirect, as in the case of various media (books, radio, television, the web, lectures and gossip) some judgement as to its reliability needs to be exercised.

Much of what is considered to be abstract thought can be performed by any self-respecting computer. At the moment this has to be programmed by a human but with rapidly advancing AI this will doubtless no longer be the case.

As far as is known, there is little or no evidence of any organism – and least of all computers – having any self-conscious awareness. In the case of computers there has been no evolution whatsoever in this respect.

In the case of living creatures, we are faced with a stark dichotomy: because of the mechanics of genetic inheritance, either all have at least some element of self-conscious awareness or there is a need for something quite separate in the human race. In the former, there can be no element of individual freedom, no understanding worth anything. In the case of man, the situation is different and the unalterable laws of physics need sometimes to be bent to his will.

Our knowledge of right and wrong is either a result only of natural evolution or at base it is innate. If it is the former, it is without content (as indeed is this writing).

---

[9] Many years ago, we were lectured on giving lectures: "Tell them what you are going to say. Say it. Tell them what you have told them." The Contents, the body of the text and this Summary exemplify the advice.

Nobody believes this.

Belief in the existence of God or this or that religion comes up against the same argument. It only has value if something apart from material evolution exists. In the case of the latter, even if there is some innate knowledge, the starting point of any belief must be down to nurture and education; its development is entirely individual and for the most part unarguable. The opening page of St. Augustine's Confessions is pertinent.

If man (after all, we say homo sapiens, not homo vel mulier sapiens) has a non-material element – we will call it soul – then, for practical reasons the moment of ensoulment will play a critical role in the abortion debate. It is the moment when a foetus becomes a person and abortion becomes homicide. A minority of women regard their foetuses as humans; a majority of states and religions do, and legislate time limits when abortions are legal. These vary from six to twenty weeks.

Although not of general practical significance, the body-soul problem remains an enduring double mystery. How can a non-material substance interact with a material one, and how is man ensouled. Both dualistic and hylomorphic[69] solutions are unsatisfactory. The former due to the impossibility of imagining how a spiritual substance could 'inhabit' a person; the latter, because although it does offer a beguiling possibility of understanding, it is close to being unintelligible!

Memory has both a material and a non-material element. After death the material element is wholly destroyed. Now, moving into realms of speculation, all memories are held complete in the soul, but a human body is needed to access them. Not any body but specifically the one whose memories they are. Hence the necessity for the resurrection of the body. Of the nature of this body nothing sensible can be said.

Finally, an argument is made, which can only have any force if there is a God, that what we sense is its ultimate reality: the falling tree in an unpopulated forest does so with a resounding crash, which man is equipped to hear.

All becomes intelligible if there is a Triune God, with Christ, of one substance with the Father and Holy Spirit, both God and man.

---

[69] Hylomorphism is the Aristotelian theory of matter and form, later refined by Aquinas

Index